КРЕМНИЙ И ЕГО БИНАРНЫЕ СИСТЕМЫ

KREMNII I EGO BINARNYE SISTEMY

SILICON AND ITS BINARY SYSTEMS

SILICON
and its
BINARY SYSTEMS

A. S. Berezhnoi

TRANSLATED FROM RUSSIAN

Springer Science+Business Media, LLC

1960

The underlying Russian text
was published by the
Academy of Sciences of the Ukrainian SSR
in Kiev in 1958

Library of Congress Catalog Card Number: 60-8724

ISBN 978-1-4899-4631-7   ISBN 978-1-4899-4629-4 (eBook)
DOI 10.1007/978-1-4899-4629-4

© *1960 Springer Science+Business Media New York*
Originally published by Consultants Bureau Enterprises, Inc. in 1960.
Softcover reprint of the hardcover 1st edition 1960

CONTENTS

PREFACE

The directives of the Twentieth Congress of the Communist Party of the Soviet Union provide for new progress in science and industry in the USSR. Ever increasing attention is being paid to the production of various materials at high temperatures and pressures. This pertains especially to metallurgy, in which technological processes are now being intensified considerably.

As a result, it is necessary to apply materials with properties which make these new processes possible. Among such materials, the heat-resistant and refractory substances, carbides, nitrides, borides, oxides, and silicides are of particular importance. In general, the silicides are the compounds which have been studied least. Numerous investigations have shown that it is possible to make practical use of a series of silicides of transition metals from Groups IV–VI as refractory materials. The preparation of electrical heat elements, operating at temperatures up to 1700°, and coatings for heat-resistant metals, especially molybdenum, are of great value not only for laboratory techniques, but also for industrial furnaces.

The solution of certain geomineralogical problems, and the practical use of silica have necessitated numerous investigations of this widely available substance and these investigations have yielded unexpected new results. These allow a new approach to the solution of a number of problems.

Investigations of carborundum, especially those by Soviet experts, have also led to extremely important results.

However, the new data are scattered in numerous publications, often in issues not readily accessible, and have therefore remained insufficiently known. More than 40 years have passed since the publication of the last review of binary compounds of silicon by Hönigschmid, but courses on the physical chemistry of silicates still do not devote enough time to the examination of these substances, apart from silica. Therefore, a critical evaluation of the accumulated data would be valuable. On this basis we undertook to

1

present the data in the form of a small monograph, which is now offered to the reader.

The author would be grateful for critical comments concerning the contents of the book and hopes that it will be useful to a large group of readers interested in silicon chemistry as it will give an idea of the present state of the problem obviating the expenditure by the reader of a lot of time in searching through material in periodicals.

The Author

INTRODUCTION

The development of new technical fields has necessitated a search for new materials for various purposes. In this connection, interest in silicon and its compounds has recently increased considerably. The exceptionally large amount of silicon in the earth's crust provides the raw material for production on an industrial scale of elementary silicon and its compounds. A series of valuable properties of these materials, especially stability at high temperatures, makes it probable that the importance of silicon and its compounds will increase in the near future.

There are extensive textbooks and manuals on the chemistry of silicon and its compounds [1, 2], especially on organosilicon compounds [3, 4, 5] and silicates [6-20]. The monographic literature covers silicides a little less thoroughly [1, 2, 21-29, 690]. Lately, the periodical literature has included a large number of papers which give new data on the properties of silicon and its compounds. These data have demonstrated the errors in some concepts which had been put forward earlier and have provided new ideas on the properties of silicon compounds, particularly of silicides of Group IV, V, and VI transition metals, which are of considerable value for practical use. The new data on silicides are of outstanding value as they show that these materials may be used in high-temperature techniques (new refractories, high-temperature, current-conducting parts, protective coatings on metals, etc.). The information that has been accumulated and is given below on the binary compounds of silicon is largely concerned with this trend.

SILICON AND ITS PROPERTIES

In the Russian language, the name of the 14th element in the Periodic System, in Row III and Group IV, is derived from the Russian word for flint, in which the element occurs, while in Western Europe, the name silicon is derived from the Latin name for flint, silex. Silicon was first isolated in a more or less pure state by Gay-Lussac and Thenard in 1811, who reduced SiF_4 with metallic potassium, but it was described in 1823 by Berzelius. In 1856, Wöhler obtained better results by isolating silicon from K_2SiF_6 by an aluminothermal method. Technically pure silicon is now prepared by reducing silica sand with carbon in the presence of iron in electric furnaces at a temperature of about 1900°. The iron reduces carbide formation, but also results in impurities in the silicon (up to 3–5% of Fe). The reaction

$$SiO_2 + 2C = Si + 2CO$$

is accompanied by heat absorption (135 kcal/mole).

Carborundum also reduces silica at very high temperatures by the reaction

$$SiO_2 + 2SiC = 3Si + 2CO.$$

Purer silicon is obtained by reducing silica with magnesium.

Wartenberg [640] proposed the removal of metallic impurities (aluminum, etc.) from silicon in the form of chlorides by the action of a stream of $SiCl_4$ vapor on the surface of the molten material. The impurity content was reduced to approximately 0.01% by this process. Boron and oxygen, however, remained in the silicon.

It is difficult to prepare highly pure silicon due to its reaction with the refractories in which the silicon is melted. Zintl et al. [30] showed that molten silicon reacted at 1450° in vacuum with a fireclay or zircon ($ZrSiO_4$) crucible, whereas its reaction with oxides gave the corresponding metal and gaseous SiO. When the reaction product (SiO vapor) was removed, the reduction of the metal oxides by silicon proceeded vigorously. Similarly, silicon reduced magnesium oxide [31]. Technical silicon may be purified by treatment with HCl and HF [240].

4

Experimental investigations of silicon in a helium atmosphere at the maximum temperature for 15 minutes showed [32, 33] that at 1400° no reaction was observed between silicon and ZrO_2, but a weak reaction occurred with MgO, Al_2O_3, and TiO_2. Under these conditions and at 1600°, silicon reacted noticeably with BeO, MgO, Al_2O_3, TiO_2, ZrO_2, and ThO_2. Mullite was formed in the reaction with Al_2O_3 and with $ZrO_2-ZrSiO_4$. Measurement of the surface tension γ and the wetting angle θ of silicon in the molten state at 1450° on various materials and the adhesion energy W gave the results presented in Table 1 [34].

Table 1

Results of Determining γ, θ, and W with Molten Silicon [34]

Ceramics	Atmosphere of hydrogen			Atmosphere of helium		
	γ	θ,degree	W_{ad}, erg/cm^2	γ	θ,degree	W_{ad}, erg/cm^2
BeO	—	88	760	—	76	560
MgO	860	101	590	740	95	675
Al_2O_3	—	95	670	—	100	580
TiO_2	—	—	—	730	107	520
ZrO_2	—	90	730	725	96	650
C (graphite)	—	0	—	—	0	—
Be_2C	—	54	1200	—	63	1100

The adhesion energy was calculated from the equation

$$W_{ad} = \gamma_{L-V}(1 + \cos \theta).$$

The wetting angle in Table 1 is given for melting without a lag time. It decreased with time.

In general, molten silicon wetted all the refractories investigated, although complete wetting occurred only in the case of graphite and, after a long period, in the case of Be_2C.

Silicon formed SiO with silica and SiC with carbon at high temperatures. Thus, Dinas and carbon refractories cannot be suitable for melting very pure silicon.

Investigations were made of the interaction of molten silicon with TiC at temperatures of 1600° and 1800° in an atmosphere of hydrogen; with TiN first with a mixture of hydrogen and nitrogen (up to 1200°) and then with helium [35]. As a result, it was found that titanium carbide was moderately corroded by silicon at 1800° to form a phase similar to carborundum. Molten silicon penetrated the pores of a titanium nitride plate (15% porosity as in the case of TiC) without noticeable corrosion but with slight solution of the silicon in the TiN.

The borides of Group IV–VI transition metals are more stable to contact with silicon than the disilicides of these metals to contact with boron. It is considered that these borides should be good materials for crucibles used in the preparation of pure silicon monocrystals [36].

The above indicates that further study on refractories for melting very pure silicon is needed.

Silicon is not found in a free state in nature. Of its binary compounds, only silica is found under natural conditions, forming about 57% of the earth's crust in the form of various silicates (of which about 13% is free silica), moissanite (SiC) in the Canyon Diabolo meteorite in Arizona, USA, and, very rarely, in sedimentary rock [37]. The possible presence of silicon halides in volcanic gases has been suggested. Silicon is also of great importance in living organisms [283].

According to Fersman [38] the relative abundance of silicon is 26.20%. The amount of silicon in the earth's crust is exceeded only by oxygen, and silicon exceeds carbon by a factor of almost 75.

The atomic weight of silicon is 28.09; according to calculations, the atomic volume of the normal form is 11.51. Three stable (non-radioactive) isotopes are known for silicon [284] with the mass numbers 28 (92.27%), 29 (4.68%), and 30 (3.05%). At the present time, three radioactive silicon isotopes are known and these have the mass numbers: 27(β^+, 2.0 Mev, half-life 4.9 sec.), 31(β^-, 1.8 Mev, half-life 2.7 hours) and 32(β^-, about 0.1 Mev, half-life not less than 100 years). The short half-lives of the first two radioactive silicon isotopes make it difficult to use them as tracers in investigating compounds of this element. Only the isotope Si^{32} may be used for this purpose.

The radius of the silicon atom is 1.33 A, its covalent radius, 1.17 A, and the radii of its ions Si^{4-} and Si^{4+}, 1.98 and 0.39 A, respectively.

The electronegativity of silicon for a hybrid bond equals 7.88 ev and is thus between that of bromine (6.41 ev) and carbon (8.30 ev), approaches that of nitrogen (with Sp^4 electrons – 7.95 ev) and is considerably exceeded by that of oxygen (with $S\bar{p}^5$ electrons – 11.38 ev). In binary compounds, silicon may be either a donor or acceptor of electrons.

According to the most generally accepted data, the melting point of silicon is 1414°, but according to the latest data [640], it is 1423°, the heat of fusion, 11.1 kcal/g-at., the boiling point 2600°, and the heat of evaporation 71 kcal/g-at. According to Ruff and Konschak [39] the vapor pressure of silicon is 1 mm Hg at a temperature of 1687° and reaches 760 mm Hg at 2442°, i.e. at a somewhat lower temperature than that indicated above. According to Brewer [40], the changes in the vapor pressure of silicon with temperature are characterized by the following data:

Temperature, °C	Silicon vapor pressure, mm Hg
1207	$7.6 \cdot 10^{-4}$
1327	$7.6 \cdot 10^{-3}$
1467	$7.6 \cdot 10^{-2}$
1647	$7.6 \cdot 10^{-1}$
1867	7.6
2477	760

These data lead to the following approximate analytical relation of P_{Si} and T:

$$\lg P_{Si} = 9.90 - \frac{19\,200}{T},$$

where P_{Si} is the vapor pressure of silicon in mm Hg and T is the absolute temperature.

Silicon usually crystallizes in a cubic system of the diamond type (type A_4, group $O_h^7 - Fd3m$), see Fig. 1. The value of the edge of the elementary cell of silicon is $a = 5.43048$ kX; $z = 8$; the distance between atoms is 2.35 kX. The calculated specific gravity is 2.44 g/cc.

For this form of silicon, the hardness equals 6.5 on the Mohs scale and 240 kg/mm^2 on the Brinell scale, whereas the microhardness equals 1808 kg/mm^2 (the microhardness of diamond is 8000 kg/mm^2). The compression strength is 947 kg/cm^2. The modulus of elasticity is 149,519 kg/mm^2 [544], the compression coefficient, $\beta = 0.325 \cdot 10^{-6}$ cm^2/kg. The heat conductivity of polycrystalline silicon at room temperature is 0.20 cal/cm · sec · deg. The heat conductivity of silicon monocrystals was studied by Hull and Geballe [372] and by White and Woods [692], who found that the value of θ for silicon is 790°K, the mean coefficient of linear thermal expansion (18–1000°) $\alpha = 3.72 \cdot 10^{-6}$ [41]. The temperature dependence of the linear expansion coefficient of silicon is as follows [50]: $\alpha 10^{-6} = 3.1528 + 1.847 \cdot 10^{-3}\ t$. According to Boltaks [42], the refractive index $N = 3.87$ (diamond – 2.417). Silicon becomes luctile above 900° [43]. When silicon is cooled rapidly, twin formation is sometimes observed [691].

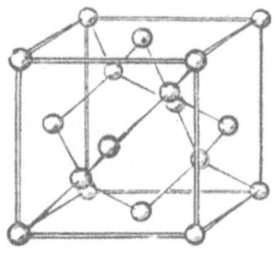

Fig. 1. Silicon crystal lattice. Type A_4 (diamond); space group $O_h^7 - Fd3m$. $a = 5.43048$ kX, $z = 8$, Si–Si distance equals 2.35 kX.

The electroconductivity of silicon is used in the production of semiconductor amplifiers and photoresistances. The temperature coefficient of electroconductivity of silicon is negative, though it is sometimes given as positive for impure silicon. The value of the electroresistance of silicon depends greatly on the purity and is highest for spectrally pure material [42]. Litton and Anderson [92] found resistance of 3–8 ohm · cm for very pure silicon prepared by thermal decomposition of SiI_4.

The reactivity of silicon depends on the method by which it is prepared. The reaction

$$3CaSi_2 + 2SbCl_3 \rightarrow 6Si + 2Sb + 3CaCl_2$$

gives silicon of very high reactivity. This is apparently explained by the retention of the "nest" disposition of the silicon atoms as found in the original $CaSi_2$ and also by the fibrous structure of the material obtained [44].

Johnson and Emick [45] prepared fibrous silicon monocrystals by reacting $SiCl_4$ vapor, diluted with argon or hydrogen, with zinc vapor at a temperature of 800–1000°. The brown silicon fibers obtained were examined with an electron microscope. These fibers contained inclusions of zinc, zinc oxide, and zinc chloride. The metal and zinc chloride were removed by evaporation in vacuum at 1000°. Each fiber, which was about 1 μ in diameter and up to 10 μ in length, was found to be a silicon monocrystal. After five months, a study of the surface of the preparation did not reveal any crystalline silicon and the authors considered this to be due to the action of electrons. Silicon which has little chemical activity, but is finely divided, is prepared by decomposing SiO at 1600–1700° [373].

Using an x-ray method, Czechoslovakian investigators [46] have recently demonstrated the existence of a noncubic, probably hexagonal, form of silicon, which is obtained by heating the cubic form to high temperatures. The new form of silicon is unstable and rapidly changes into the cubic form, but this depends on the method of preparing the sample, the size of the crystals, and the pressure. Briquets of silicon and about 20% CaF_2 and 1% NaF or KF were heated to 900° for two hours and then rapidly quenched in water. It was then established that the addition of fluorides promoted an increase in the crystal sizes and stabilized the new form of silicon at room temperature.

The x-ray powder pictures of the new form of silicon showed seven lines. which could be indexed in a hexagonal system with the following lattice parameters: $a = 6.86$ A, $c = 10.29$ A.

X-ray powder pictures of silicon, taken immediately after the latter had been evaporated in vacuum and its vapor condensed on cold glass, showed two of these new lines. After four months, this preparation had only the lines of the cubic modification of silicon. A high-temperature picture of a sample of silicon powder, heated to 700°, contained some of the new lines. These lines disappeared from the x-ray powder pictures when the preparation was cooled or heated to 900°. On the contrary, the number of new lines on the high-temperature x-ray powder pictures of briquet silicon increased when the temperature was raised from 700 to 900° and these lines did not disappear on cooling.

These experiments, which were of a preliminary nature, seem to indicate the existence of a hexagonal modification of silicon similar to the graphite form of carbon. However, this silicon modification is apparently very unstable.

In contrast to the results of investigations carried out by Heyd [46], Wood [642] did not obtain any data on the existance of a noncubic form of silicon. Only lines corresponding to the normal cubic modification were found on the x-ray pictures of "highly pure" silicon, taken at temperatures of 700, 800, and 900°. Similar results were obtained on investigating silicon isolated from $SiCl_4$ vapor and hydrogen. Thermograms of very pure silicon at 1000° did not show any effects that would indicate a modificational change. From this Wood concluded that the high-temperature form of silicon described by Heyd did not actually exist.

As Wood did not exactly repeat Heyd's experiments and Heyd has not as yet published the complete results of the investigations, the problem of whether or not a noncubic modification of silicon exists remains unsolved.

Latest investigations have shown that "amorphous" silicon does not exist, but is a finely crystalline form of the cubic modification.

Silicon reacts readily with alkalies, including very dilute aqueous solutions of them obtained, for example, by contact with alkali-containing glass. With the exception of HF, acids do not react with silicon. Law and François [374] studied gas absorption on silicon and Dunlap [375] studied the diffusion of impurities in it.

In contrast to carbon, silicon forms relatively unstable chains and this is connected with the lower bond energy [5]:

Bond	Bond energy, kcal/mole
Si–Si	42.5
Si–C	58.0
C–C	62.8

The Si–Si bond is homeopolar. The longest known chain of silicon atoms in compounds that have been investigated is in $Si_{25}Cl_{52}$ and includes 25 units [74]. Polysilanes up to $Si_{14}H_{30}$ were obtained with great difficulty, but they could only be investigated up to Si_6H_{14} due to the low stability of polysilanes with a large number of silicon atoms [5].

Very pure silicon, like germanium, is used as a semiconductor. Technically pure silicon is also of great importance and the best type (KRO) must fulfill the following requirements as regards chemical composition [544]: not less than 99.0% Si, not more than 0.5% Fe, not more than 0.5% Al, not more than 0.5% Ca, and not more than 1% of impurities. Silicon with 98, 97, and 95.5% Si content is also produced. Technically pure silicon is used for the preparation of nonferrous alloys (bronze silicide and Silumin) and sometimes as a reducing agent.

The preparation of refractory fibers from silicon with silicide bonding has been recommended [669].

PREPARATION OF BINARY SILICON COMPOUNDS

The most important methods of preparing binary silicon compounds are the following.

Direct reaction of silicon with the other element or its hydride

$$Me + Si \rightarrow MeSi,$$

$$MeH + Si \rightarrow MeSi + H_2.$$

Reduction of a metal oxide with silicon

$$MeO + Si \rightarrow MeSi + SiO_2.$$

Reduction of a mixture of metal oxide and silica with carbon

$$MeO + SiO_2 + C \rightarrow MeSi + CO.$$

Aluminothermy and similar methods

$$MeO + Al(Mg) + SiO_2 + S \rightarrow MeSi + \text{slag containing Al(Mg) and S.}$$

Preparation by exchange reactions, including those involving the gas phase

$$(Cu - Si) + Me \rightarrow MeSi + Cu,$$

$$Mg_2Si + 4HCl \rightarrow SiH_4 + 2MgCl_2,$$

$$Me + SiCl_4 + H_2 \rightarrow MeSi + HCl.$$

Electrolysis from a flux melt

$$K_2SiF_6 + MeO \rightarrow MeSi + KF.$$

Compounds of silicon with hydrogen of the Si_nH_{2n+2} series (silanes) are prepared by decomposition of magnesium, calcium, and lithium silicides with acids and also, for example, decomposition of triethyl silicoorthoformate

$$4SiH(OC_2H_5)_3 = 3Si(OC_2H_5)_4 + SiH_4.$$

The decomposition of magnesium silicide, Mg_2Si, with acids is a complex process. According to Schwarz [208], it is summarized by the reactions

$$Mg_2Si + 2H_2O = 2MgO + H_2 + (SiH_2),$$

$$(SiH_2) + H_2O = (SiH_2O) + \text{silanes.}$$

Prosilane (SiH_2O) reacts with water to give SiO_2 and hydrogen

$$SiH_2O + H_2O = SiO_2 + 2H_2.$$

SiO_2 and H_2 are also obtained in small amounts from unstable silene

$$(SiH_2) + 2H_2O = SiO_2 + 3H_2.$$

It is also possible that the hydrolysis product of magnesium silicide $(OH)_2Mg_2SiH_2$ [209] is also partially formed and this decomposes to form silanes, hydrogen, and SiO_2.

With an increase in molecular weight, the amount of polysilanes in the mixture decreases. The silanes are separated into discrete substances by fractional distillation at very low temperatures in the absence of air.

According to Stock and Somieski [51], a polymer of the silene series $(SiH_2)_x$ is formed by the reaction of dichlorosilane and sodium amalgam; for example

$$SiH_2Cl_2 + 4Na = H_2SiNa_2 + 2NaCl,$$

$$(H_2SiNa_2 + Hg)_x = (SH_2)_x + X\, HgNa_2.$$

According to Schwarz [47], polymerized silene is formed by decomposing silanes (for example Si_3H_8) with a silent discharge and also by decomposing CaSi with glacial acetic acid or an alcohol solution of HCl.

According to Dolgov [52], polymers of the $(SiH)_x$ type are formed by decomposing silicides and polysilanes.

The direct reaction of silicon with another element, described by Moissan in the last century, is very valuable for preparing numerous binary silicon compounds. This method has several variants: reaction in the solid phase, including simultaneous pressing (hot pressing), fusion of the components, and reaction of a metal with silicon vapor. The reaction usually proceeds at high temperatures. The reactions forming binary silicon compounds proceed at a noticeable rate at the following temperatures:

Reacting element	Temperature of reaction with silicon, °C
F	20
Cl	500
O	400
S	600
N	1100
Mg	600
Zr	940
C	1100
Pt	below 1000

Silicides of Group IV, V, and VI transition metals are usually prepared by hot pressing of fine grained mixtures of silicon and metal in graphite molds at high temperatures, often up to 1600–1800° and even higher for tungsten and tantalum silicides. The pressure used is 700 kg/cm^2 [48, 49, 490].

The use of hydrides of transition metals makes the synthesis of silicides of the latter easier. The formation reactions of these metal silicides are usually exothermal. To obtain pure silicides, it is necessary to synthesize them in an atmosphere of an inert gas (usually argon) or in vacuum. Silicides react slowly with carbon and oxides. Thus, in a number of cases, the synthesis may be carried out in graphite crucibles or vessels of highly refractory oxides (Al_2O_3, BeO, and others). Wallbaum [484] obtained vanadium, niobium, and tantalum silicides by sintering powders of these metals with silicon in alumina crucibles in an argon atmosphere. Silicides may also be prepared by reaction with gaseous silicon and apparatus made from tungsten or thorium dioxide is suitable for evaporating the silicon [291].

The preparation of binary silicon compounds and their properties are examined in greater detail in the sections describing the systems. Data on a series of properties of binary silicon compounds are given in Table 2. The majority of the binary silicon compounds are not transparent. Therefore, no optical properties are given in Table 2. In the cases where they have been determined, the text gives appropriate data.

The volume changes during the formation of binary compounds are given in Table 2 as calculated for the cubic modification of silicon and the stable crystalline form of the other element. These data are given for a temperature of 20° without taking into account thermal expansion. The crystal lattice parameters are given mostly in A and, more rarely, in kX units. As the difference in these units was often less than the accuracy of the measurements of the crystal lattice parameters, it would not have been worthwhile recalculating the values for one system of measurement. Wherever possible, the specific gravity was calculated from the value of the elementary cell. What has been said above regarding the units of measurement for the elementary cell should be taken into account in evaluating the accuracy of the values for specific gravity and molecular volumes. However, these inaccuracies are insignificant for practical purposes.

TABLE 2

The Most Important Properties of Silicon

Sample No.	Chemical formula	Molecular weight	Molec- ular volume	Change in volume during formation from the elements, ΔV, %	Crystal system	Space group	Crystal lattice		
							a	b	c
1	Si	28,09	11,51	—	Cubic	O_h^7	5,4305	—	—
	Si	28,09	—	—	Hexag- onal	—	6,86	—	10,29
2	SiH_4	32,12	—	—	Cubic	—	—	—	—
3	Si_2H_6	62,23	—	—	—	—	—	—	—
4	Si_3H_8	92,33	—	—	—	—	—	—	—
5	Si_4H_{10}	122,44	—	—	—	—	—	—	—
6	$Si_{14}H_{30}$	423,51	—	—	—	—	—	—	—
7	$(SiH_3)_x$	$30,11 \cdot x$	—	—	—	—	—	—	—
8	$(SiH)_x$	$29,10 \cdot x$	—	—	—	—	—	—	—
9	SiF_4	100,09	47,21	—	Cubic	T_d^3	5,41	—	—
10	Si_2F_6	170,18	—	—	—	—	—	—	—
11	$SiCl_4$	169,92	—	—	Cubic	—	—	—	—
12	Si_2Cl_6	268,92	—	—	—	—	—	—	—
13	Si_3Cl_8	367,93	—	—	—	—	—	—	—
14	Si_4Cl_{10}	466,93	—	—	—	—	—	—	—
15	Si_5Cl_{12}	565,93	—	—	—	—	—	—	—
16	Si_6Cl_{14}	664,94	—	—	—	—	—	—	—
17	$Si_{10}Cl_{22}$	1060,95	—	—	—	—	—	—	—
18	$Si_{25}Cl_{52}$	2546,01	—	—	—	—	—	—	—
19	$SiBr_4$	347,75	123,67	From liquid Br+7,9	Cubic	—	—	—	—
20	Si_2Br_6	535,68	—	—	Rhombic	—	—	—	—
21	Si_3Br_8	723,60	—	—	—	—	—	—	—
22	Si_4Br_{10}	911,52	—	—	—	—	—	—	—

and Its Binary Compounds

constants		Micro-hardness, kg/mm²	Specific gravity, g/cm³	Melting point, °C	Silicon content by weight, %	Remarks
Angles	z					
—	8	1808	2,44	14!4	100,00	Boiling point 2600°
—	—	—	—	—	100,00	The existence of this form of silicon is questioned [642]
—	—	—	Liquid 0,58	—184,7	87,45	$\Delta H°_{298°} = -14,8; \Delta F°_{298} = -9,4$ kcal/mole Boiling point —112°
—	—	—	Liquid 0,686	—132	90,28	The same—44°
—	—	—	Liquid 0,743	—117	91,27	» » +53°
—	—	—	Liquid 0,825	— 90	91,77	» » +109°
—	—	—	—	—	92,86	—
—	—	—	—	—	93,29	—
—	—	—	—	—	96,53	—
—	2	—	2,12	—95,7	28,06	$\Delta H°_{298°} = -370; \Delta F°_{298} = -360,0$ kcal/mole Boiling point —65°
—	—	—	—	—18,8	33,01	—
—	—	—	Liquid 1,483	—67,7	16,53	$\Delta H°_{298°} = -145,73; \Delta F°_{298°} = -136,2$ kcal/mole Boiling point +57,6°
—	—	—	Liquid 1,58	—1	20,89	The same+147°
—	—	—	—	67	22,90	» » +216°
—	—	—	—	—	24,06	» » +150°
—	—	—	—	—	24,82	» » +150° (15 mm)
—	—	—	—	170	25,35	Sublimes at a temperature of +200°
—	—	—	—	With decomposition	26,48	—
—	—	—	—	(With decomposition)	27,58	—
—	—	—	Liquid 2,812	5	8,08	$\Delta H°_{298°} = -95,1$ kcal/mole Boiling point 153,4°
—	—	—	—	65	10,49	Boiling point 240°
—	—	—	—	133	11,65	—
—	—	—	—	185	12,33	—

Table 2 (continued)

Sample No.	Chemical formula	Molecular weight	Molecular volume	Change in volume during formation from the elements, ΔV, %	Crystal system	Space group	Crystal lattice		
							a	b	c
23	SiI_4	535,77	130,04	+14,2	Cubic	T_h^6	11,99	—	—
24	Si_2I_6	817,70	—	—	Hexagonal	—	—	—	—
25	SiI_2	281,93	—	—	—	—	—	—	—
26	$Li_{15}Si_4$	216,46	—	—	—	—	—	—	—
27	Li_2Si	41,97	37,47	—0,2	—	—	—	—	—
28	NaSi	51,09	—	—	—	—	—	—	—
29	$NaSi_2$	80,18	60,74	+32,6	Tetragonal	—	4,97	—	16,7
30	KSi	67,19	—	—	—	—	—	—	—
31	KSi_8?	263,84	—	—	—	—	—	—	—
32	RbSi	113,53	—	—	—	—	—	—	—
33	$RbSi_8$?	310,16	—	—	—	—	—	—	—
34	CsSi	161,00	—	—	—	—	—	—	—
35	$CsSi_8$?	357,63	—	—	—	—	—	—	—
36	Cu_6Si	410,51	—	—	Hexagonal	D_{6h}^4	2,59	—	4,18
37	Cu_5Si	345,94	(43,8)	(—7,0)	Cubic	$O^{6,7}$	6,211	—	—
38	~Cu_4Si	282,37	(37,6)	(—6,0)	—	—	—	—	—
39	$Cu_{15}Si_4$	1065,91	137,01	+43,4	Cubic	T_d^6	9,694	—	—
40	Cu_3Si	218,80	—	—	Hexagonal	—	—	—	—
41	Mg_2Si	76,73	38,36	—2,5	Cubic	O_{13}^5	6,338	—	—
42	Ca_2Si	108,25	49,88	(α-Ca) —21,1	Rhombic	D_{2h}^{16}	9,002	7,667	4,799
43	CaSi	68,17	21,24	(α-Ca) —43,2	Rhombic	D_{2h}^{17}	3,91	4,59	10,795
44	$CaSi_2$	96,26	38,50	(α-Ca) —21,4	Trigonal	D_{3d}^5	10,4	—	—

constants		Micro-hardness, kg/mm²	Specific gravity, g/cm³	Melting point, °C	Silicon content by weight, %	Remarks
Angles	z					
—	8	—	4,12	120,5	5,24	$\Delta H_{298°} = -31,6$ kcal/mole Boiling point 290°
—	—	—	—	250	6,87	—
—	—	—	—	—	9,96	—
—	—	—	—	720	51,90	Silvery gray
—	—	—	1,12	Decomposes on heating	66,93	Dark blue-violet, very hygroscopic
—	—	—	—	—	54,98	Needles with metallic luster
—	4	—	1,32	—	70,06	The same
—	—	—	—	—	41,81	Dark, lustrous solid mass
—	—	—	—	—	85,17	—
—	—	—	—	—	24,74	Dark
—	—	—	—	—	72,45	—
—	—	—	—	—	17,45	—
—	—	—	—	—	62,84	—
—	—	—	—	(Incongruent) 852	6,84	β-Phase of the Cu—Si system
—	20 atoms	—	7,9	(Decomposes in the solid phase) 716	8,12	γ-Phase of the Cu—Si system, β-Mn type
—	—	—	(7,5)	(Incongruent) 824	9,95	δ-Phase of the Cu—Si system, deformed cubic lattice
—	4	—	7,78	(Decomposes in the solid phase) 800	10,54	Or $Cu_{31}Si_8$ (10.2% Si), molecular weight= 2195.15, ε-phase of the Cu—Si system, $D8_6$ type lattice according to Arrhenius
—	—	400	—	850	12,84	η-Phase of the Cu—Si system according to Arrhenius
—	4	450	2,00	1085	36,62	$\Delta H°_{form.} = -18,5 \pm 1,5$ kcal/mole $C_p = = 15,4 + 4,15 \cdot 10^{-3} T$
—	4	—	2,17	(Incongruent) 910	25,95	According to other data— cubic $a=4,73$, $\Delta H°_{form.} = -50 \pm \pm 3$ kcal/mole
—	4	—	3,21	1245	41,21	Homeopolar bond, $\Delta H°_{form.} = 36 \pm 2$ kcal/mole
(α) 21°30′	2	—	2,5	(Incongruent) 1020	58,36	$\Delta H°_{form.} = -36 \pm \pm 2$ kcal/mole

Table 2 (continued)

Sample No.	Chemical formula	Molecular weight	Molecular volume	Change in volume during formation from the elements, ΔV, %	Crystal system	Space group	Crystal lattice		
							a	b	c
45	SrSi	115,72	—	—	—	—	—	—	—
46	SrSi$_2$	143,81	—	—	—	—	—	—	—
47	BaSi	165,45	—	—	—	—	—	—	—
48	BaSi$_2$	193,54	—	—	—	—	—	—	—
49	BaSi$_3$	221,63	—	—	—	—	—	—	—
50	Ba$_2$Si$_7$?	471,35	—	—	—	—	—	—	—
51	BaSi$_4$?	249,72	—	—	—	—	—	—	—
52	B$_6$Si	93,01	42,34	+36,5	Cubic	O_h^1	4,142	—	—
53	B$_3$Si	60,55	24,03	+13,0	Tetragonal	—	2,829	—	4,765
54	SiC	40,09	12,49	—25,6	Cubic	T_d^2	4,352	—	—
	»	40,09	12,50	—25,6	Hexagonal	C_{6v}^4	3,079	—	10,254
	»	40,09	12,50	—25,6	»	C_{6v}^4	3,082	—	15,118
	«	40,09	12,50	—25,6	»	C_{6v}^4	3,073	—	20,106
	»	40,09	12,50	—25,6	»	C_{6v}^4	3,073	—	25,133
	»	40,09	12,50	—25,6	»	C_{3v}^5	3,079	—	37,78
	»	40,09	12,50	—25,6	»	—	3,07	—	(~43)
	»	40,09	12,50	—25,6	»	—	3,073	—	47,75
	»	40,09	12,50	—25,6	»	C_{3v}^5	3,079	—	52,88
	»	40,09	12,50	—25,6	»	—	3,073	—	67,86
	»	40,09	12,50	—25,6	»	C_{3v}^5	3,079	—	83,10
	»	40,09	12,50	—25,6	»	C_{3v}^5	3,079	—	128,43
	»	40,09	12,50	—25,6	»	—	3,073	—	188,50
	»	40,09	12,50	—25,6	»	—	3,073	—	211,12
	»	40,09	12,50	—25,6	»	C_{3v}^5	3,079	—	219,09
	»	40,09	12,50	—25,6	»	C_{3v}^5	3,073	—	354,33
	»	40,09	12,50	—25,6	»	—	3,07	—	(~670)
	»	40,09	12,50	—25,6	»	—	3,07	—	(~686)
	»	40,09	12,50	—25,6	»	—	3,073	—	~730
	»	40,09	12,50	—25,6	»	C_{3v}^5	3,073	—	987,61

constants		Micro-hardness, kg/mm^2	Specific gravity, g/cm^3	Melting point, °C	Silicon content by weight, %	Remarks
Angles	z					
—	—	—	—	—	24,27	$\Delta H^{\circ}_{form.} = -113$ kcal/mole
—	—	—	—	—	39,07	$\Delta H^{\circ}_{form.} =$
						$= -147,4$ kcal/mole
—	—	—	—	—	16,98	$\Delta H^{\circ}_{form.} = -181$ kcal/mole
—	—	—	—	—	29,03	—
—	—	—	—	—	38,02	$\Delta H^{\circ}_{form.} =$
						$= -399,2$ kcal/mole
—	—	—	—	—	41,72	—
—	—	—	—	—	44,99	—
—	1	—	2,18	—	30,43	Black in color, $\alpha = 5,9 \cdot 10^{-6}$
—	1	5352	2,52	—	46,39	Black in color
—	4	—	3,210	(Conversion to α-SiC) 2100	70,07	β-Form [156, 159]
—	4	—	3,21		70,07	α-Form, type III, $4H$ [156], $\Delta H = -26,46$ kcal/mole $\Delta F^{\circ} = -26,7$ kcal/mole
—	6	3340	3,208	(Decomposes) 2700	70,07	α-Form, type II, $6H$ [156]
—	8	—	3,21	—	70,07	» » type VIII, $8H$ [170, 171]
—	10	—	3,21	—	70,07	The same $10H$ [619]
—	15	—	3,21	—	70,07	» » type I $+5R$ [156]
—	17	—	3,21	—	70,07	» »· [163]
—	19	—	3,21	—	70,07	» » [172]
—	21	—	3,21	—	70,07	» » type IV, $21R$ [156, 161]
—	27	—	3,21	—	70,07	» » $27R$ [171]
—	33	—	3,21	—	70,07	» » type VI, $33R$ [161, 166]
—	51	—	3,21	—	70,07	» » type V, $51R$ [156, 163, 171]
—	75	—	3,21	—	70,07	» » $75R$ [171]
—	84	—	3,21	—	70,07	» » $84R$ [171]
—	87	—	3,21	—	70,07	» » type VII, $87R$ [165]
—	141	—	3,21	—	70,07	» » $141R$ [174]
	267		3,21		70,07	» » [177]
—	273		3,21	—	70,07	α-Form [177]
—	297*	—	3,21	—	70,07	α-Form*. Previously [167] considered 594 erroneously [175]
—	393	—	3,21	—	70,07	α-Form, $393R$ [174]

Table 2 (continued)

Sample No.	Chemical formula	Molecular weight	Molecular volume	Change in volume during formation from the elements, ΔV, %	Crystal system	Space group	Crystal lattice		
							a	b	c
55	SiN_2	56,09	—	—	—	—	—	—	—
56	Si_2N_3	98,20	26,97	In relation to Si)+17,2	—	—	—	—	—
57	Si_3N_4	140,30	44,54	In relation to Si)+29,0	Rhombic	—	13,38	8,60	7,74
58	$(Si_2N_2)x$	84,20·x	26,56·x	In relation to Si)+11,0	—	—	—	—	—
59	SiP	59,11	—	—	—	—	—	—	—
60	$SiAs_2$	177,91	41,37	+9,9	—	—	—	—	—
61	SiAs	103,00	27,39	+11,5	—	—	—	—	—
62	SiO	44,09	19,77	(from gaseous oxygen) +71,7	Cubic	T_h^6	6,4	—	—
63	Si_2O_3	104,18	—	—	—	—	—	—	—
64	SiO_2	60,09	30,50	—	Rhombic	D_{2h}^{26}	4,72	5,16	8,36
	SiO_2 β-quartz	60,09	22,68	—	Hexagonal	$D_3^{4,6}$	4,903	—	5,394
	¡SiO_2, α-quartz	60,09	22,84	—	»	$D_6^{5,4}$	5,01	—	5,47
	SiO_2, γ-tridymite	60,09	26,01	—	Rhombic	D_{2h}^{23}	9,91	17,18	16,3
	SiO_2, β-tridymite	60,09	26,13	—	Hexagonal	—	—	—	—
	SiO_2, α-tridymite	60,09	26,95	—	»	D_{6h}^4	5,03	—	8,22
	SiO_2, β-cristobalite	60,09	25,68	—	Tetragonal	$D_4^{4,8}$	4,96	—	6,92
	SiO_2, α-cristobalite	60,09	27,07	—	Cubic	O_h^7	7,12	—	—

constants		Micro-hardness, kg/mm^2	Specific gravity, g/cm^3	Melting point, °C	Silicon content by weight, %	Remarks
Angles	z					
—	—	—	—	—	50,08	Silicocyanamide
—	—	—	3,64	—	57,21	Specific gravity according to Weiss and Engelhardt (1910)
—	12	3350	3,15	(Dissociation) 1900	60,06	$\Delta H°_{298} = -179,3$; $\Delta F = -154,7$ kcal/mole
—	—	—	3,17	—	66,72	Silicocyanogen
—	—	—	—	(Dissociation) 1140	47,52	—
—	—	—	4,30	(Incongruent 944	15,79	Coarsely crystalline gray-black mass
—	—	—	3,76	1083	27,27	Platelike structure
—	8	—	2,23	Above 1700	63,71	—
						—
—	—	—	—	1635	53,93	—
—	4	—	1,97	1420	46,75	Fibrous modification
—	3	820	2,65	(Conversion into α) 573	46,75	—
—	3	—	2,52	(Conversion into α-tridymite) 870	46,75	—
—	64	—	2,31	(Conversion into β) 117	46,72	—
—	—	—	(2,30)	(Conversion into α) 163	46,72	There are indications that there are conversions of tridymite at 210 and 475°
—	4	—	2,23	(Conversion into α-cristobalite) 1470	46,72	—
—	4	—	2,34	(Conversion into α-cristobalite) 218	46,72	The β-cristobalite is occasionally indexed in a rhombic or triclinic system, which is unnecessary
—	8	—	2,22	1723	46,72	—

Table 2 (continued)

Sample No.	Chemical formula	Molecular weight	Molecular volume	Change in volume during formation from the elements, ΔV, %	Crystal system	Space group	Crystal lattice		
							a	b	c
64	SiO_2, coesite	60,09	19,96	—	Monoclinic	C_{2h}^6—$C2/c$	7,23	12,52	7,23
	SiO_2, keatite	60,09	24,04	—	Tetragonal	—	7,46	—	8,59
65	SiS_2	92,21	45,65	+ 4,8	Rhombic	D_{2h}^{26}	5,60	5,53	9,55
66	SiS	60,15	32,46	+17,8	—	—	—	—	—
67	$SiSe_2$	186,01	50,96	+15,5	Rhombic	D_{2h}^{26}	6,03	5,76	9,76
68	SiSe	107,05	—	—	—	—	—	—	—
69	$SiTe_2$	283,31	64,54	+23,2	Hexagonal	D_{3d}^3	4,28	—	6,71
70	SiTe	155,70	39,62	—	Cubic	—	—	—	—
71	Ti_5Si_3	323,77	75,12	(β-Ti) —15,9	Hexagonal	D_{6h}^3	7,465	—	5,162
72	TiSi	75,99	17,51	(β-Ti) —22,1	Rhombic	D_{2h}^1	3,611	4,960	6,479
73	$TiSi_2$	104,08	23,71	(β-Ti) —30,2	»	D_{2h}^{24}	8,236	4,773	8,523
	»	104,08	27,04	(β-Ti) —20,4	»	D_{2h}^{17}	3,62	13,76	3,605
74	Zr_2Si	208,53	34,76	(β-Zr) —13,4	Tetragonal	D_{4h}^{18}	6,568	—	5,360
75	Zr_5Si_3	540,37	89,91	(β-Zr) —15,2	Hexagonal	D_{6h}^3	7,870	—	5,547
76	Zr_3Si_2	329,84	—	—	Tetragonal	D_{4h}^5	—	—	—
77	Zr_6Si_5	687,77	—	—	—	—	—	—	—
78	ZrSi	117,31	20,95	(β-Zr) —18,8	Rhombic	D_{2h}^{16}	6,968	3,778	5,291
79	$ZrSi_2$	145,40	30,10	(β-Zr) —20,4	·	D_{2h}^{17}	3,71	14,73	3,66
80	HfSi	206,69	19,26	Fromα-Hf —25,0	Hexagonal	—	(6,86)	—	(12,60)
81	$HfSi_2$	234,78	29,20	(α-Hf) —21,5	Rhombic	D_{2h}^{17}	3,67	14,56	3,64
82	V_3Si	180,94	31,58	—14,6	Cubic	O_h^3	4,712	—	—

constants		Micro-hardness, kg/mm^2	Specific gravity, g/cm^3	Melting point, °C	Silicon content by weight, %	Remarks
Angles	z					
β: 120°	16?	1200	3,01	—	46,72	—
—	12	—	2,50	—	46,72	—
—	4	—	2,02	1090	30,46	White in color. Boiling point 1130°
—	—	—	1,853	(B.p.) 940	46,70	Yellow needles
—	1	—	3,65	1060 (In vacuum)	15,10	Gives a glass with a specific gravity of 2.95
--	—	—	—	(Sublimes)	26,24	—
—	2	—	4,39	1220	9,91	A fibrous variety also exists
—	—	—	3,93	898	18,04	—
—	2	986	4,31	2120	26,03	—
—	8	1039	4,34	(Incongruent) 1760	36,99	Structure according to [716]
—	8	870	4,39	About 1540	53,98	Iron gray color, electrical resistance 123 μohm·cm
—	4	890	3,85	—	53,98	On converting to the previous form $\Delta V = {}=-12,3\%$
--	4	1230*	6,00	(Incongruent) 2110	13,47	•50 g load
—	2	1340*	6,01	2250	15,59	•The same; exists only when stabilized with C, N_2, B, or O_2
—	—	—	—	—	17,03	—
—	—	—	—	—	20,42	—
—	4	1100*	5,60	(Incongruent) 2005	23,95	•50 g load
—	4	840	4,83	(Incongruent) 1520	38,64	Electrical resistance 161 μohm·cm
—	16	—	10,73	—	13,59	Apparently a rhombic structure, as for ZrSi, pseudohexagonal
—	4	865	8,04	—	23,93	—
—	2	1500 (Cuts glass readily)	5,73	2065	15,52	—

Table 2 (continued)

Sample No.	Chemical formula	Molecular weight	Molecular volume	Change in volume during formation from the elements, ΔV, %	Crystal system	Space group	Crystal lattice		
							a	b	c
83	V_2Si	129,99	23,72	—16,7	—	—	—	—	—
84	V_5Si_3	339,02	64,33	—16,4	Hexagonal	D_{6h}^3	7,121	—	4,832
	»	339,02	63,26	—17,9	Tetragonal	D_{4h}^{18}	9,410	—	4,747
85	VSi_2	107,13	22,74	—27,0	Hexagonal	D_{6h}^4	4,562	—	6,359
86	Nb_4Si	399,73	—	—	»	—	—	—	—
87	Nb_2Si	213,91	27,60	—16,8	—	—	—	—	--
88	Nb_5Si_3	548,82	77,85	—12,2	Hexagonal	D_{6h}^3	7,521	—	5,238
	»	548,82	76,22	—13,2	Tetragonal	D_{4h}^{18}	9,998	—	5,067
	»	548,82	76,86	—13,6	»	D_{2d}^{11}	6,557	—	11,860
89	$NbSi_2$	149,09	28,18	—16,8	Hexagonal	D_6^4	4,785	—	6,576
90	$Ta_{4,5}Si$	844,39	65,66	+ 8,2	»	—	6,093	—	4,909
91	Ta_2Si	390,89	28,87	13,4	Tetragonal	D_{4h}^{18} —	6,145	—	5,029
92	Ta_5Si_3	935,15	71,60	—19,7	Hexagonal	D_{6h}^3	7,459	—	5,215
	»	935,15	73,87	—17,2	Tetragonal	D_{4h}^{18}	9,86	—	5,05
93	» $TaSi_2$	935,15 237,58	75,78 26,91	—16,1 —20,7	» Hexagonal	— D_6^4	6,503 4,773	— --	11,849 6,552
94	Cr_3Si	184,12	28,24	—15,4	Cubic	O_h^3	4,555	—	—

constants		Micro-hardness, kg/mm^2	Specific gravity, g/cm^3	Melting point, °C	Silicon content by weight, %	Remarks
Angles	z					
—	—	—	5,48	—	21,61	The existence of this compound is doubtful. Perhaps this is V$_5$Si$_3$?
—	2	1430	5,27	—	24,86	Stabilized by carbon, boron, or nitrogen
—	4	—	5,36	2165	24,86	—
—	3	1090	4,71	1700	52,44	Electrical resistance 9.5 μohm·cm
—	—	—	—	(Incongruent) 1950	7,03	—
—	—	—	7,75	—	13,18	There are indications that a second β modification exists with a specific gravity of 7.34, but evidently this is Nb$_5$Si$_3$; perhaps Nb$_2$Si does not exist [681]
—	2	—	7,05	2480	15,35	Stabilized by carbon, nitrogen, and boron. Does not exist with pure niobium
—	4	500	7,20	—	15,35	Low-temperature form
—	4	—	7,14	—	15,35	High-temperature form
—	3	1050	5,29	1930	37,68	Electrical resistance of about 6.3 μohm·cm
—	2	1100	12,86	~2510	3,33	May be written as Ta$_3$(Ta$_{0,28}$Si$_{0,72}$)
—	4	1350	13,54	(Incongruent) About 2450	7,19	It is possible that this is one of the forms of Ta$_5$Si$_3$
—	2	1350	13,06	About 2500	9,01	Stabilized by carbon, nitrogen, and boron. Does not exist with pure tantalum.
—	4	—	12,66	—	9,01	—
—	4	—	12,34	—	9,01	—
—	3	1560	8,83	2400 (2200)	23,65	Electrical resistance of about 8.5 μohm·cm
—	2	—	6,52	1710	15,26	Electrical resistance 0.05 μohm·cm

Table 2 (continued)

Sample No.	Chemical formula	Molecular weight	Molecular volume	Change in volume during formation from the elements, ΔV, %	Crystal system	Space group	Crystal lattice		
							a	b	c
95	Cr_5Si_3	344,32	58,96	—17,0	Tetragonal	D_{4h}^{18}	9,16	—	4,64
	»	344,32	60,30	—15,1	Hexagonal	D_{6h}^{3}	6,979	--	4,716
96	CrSi	80,10	14,78	—21,4	Cubic	T^4	4,607	—	—
97	$CrSi_2$	108,19	24,59	—18,8	Hexagonal	D_6^4	4,422	—	6,351
98	Mo_3Si	315,94	35,34	—11,0	Cubic	O_h^3	4,890	—	—
99	Mo_5Si_3	564,02	69,63	—14,6	Tetragonal	D_{4h}^{18}	9,82	—	4,97
	»	564,02	69,04	—15,3	Hexagonal	D_{6h}^{3}	7,271	—	4,992
100	$MoSi_2$	152,13	24,86	—23,4	Tetragonal	D_{4h}^{17}	3,197	—	7,871
101	W_3Si	579,85	—	—	Cubic	O_h^3	—	—	—
102	W_5Si_3	1003,87	67,56	—17,7	Tetragonal	D_{4h}^{18}	9,54	—	4,93
	»	1003,87	65,19	—20,6	Hexagonal	D_{6h}^{3}	7,18	—	4,84
103	WSi_2	240,10	24,50	—24,7	Tetragonal	D_{4h}^{17}	3,212	—	7,880
104	Mn_3Si	192,88	27,63	(γ-Mn) - 19,5	Cubic	O_h^9	2,85	—	—
105	Mn_5Si_3	358,92	59,72	(γ-Mn) —17,7	Hexagonal	D_{6h}^{3}	6,898	—	4,802
106	MnSi	·85,02	14,22	(γ-Mn) —25,7	Cubic	T^4	4,548	—	—

constants		Micro-hardness, kg/mm^2	Specific gravity, g/cm^3	Melting point, °C	Silicon content by weight, %	Remarks
Angles	z					
—	4	—	5,84	1600	24,47	Sometimes written as Cr$_3$Si$_2$ (26.45% Si). Electrical resistance 0.7 μohm·cm
—	2	—	5,71	—	24,47	Stabilized by carbon, nitrogen, or oxygen
—	4	—	5,42	1545	35,07	Electrical resistance 0.25 μohm·cm
—	3	1150	4,40	1550	51,93	Electrical resistance 7 μohm·cm
—	2	1310	8,94	(Incongru-ent) 2050	8,89	—
—	4	1170	8,10	2100	14,94	Sometimes written as Mo$_3$Si$_2$ (16.32% Si). The hardness is given for this composition
—	2	—	8,17	—	14,94	Stabilized by carbon
—	2	1260	6,12	2030 (1870)	36,93	Relative to Mo when Si vapor is formed $\Delta V = 260\%$. Electrical resistance 21.5 μohm·cm; $\Delta H_{298°} = -16$ kcal/g-at. Si
—	2	—	—	—	4,84	The existence of this silicide has not been established definitely
—	4	770	14,86	2320	8,39	More often written as W$_3$Si$_2$ (9.23% Si). The hardness is given for this composition
—	2	—	15,40	—	8,39	Stabilized by carbon
—	2	1090	9,80	2165	23,40	Electrical resistance 33.4 μohm·cm. Melting point 2050° according to Chervinko
—	—	—	6,98	(Incongru-ent) 1075	14,56	Statistical distribution of atoms in the α-Fe structure
—	2	—	6,01	1285	23,48	—
—	4	8.5 on the Mohs scale	5,98	1275	33,04	—

Table 2 (continued)

Sample No.	Chemical formula	Molecular weight	Molecular volume	Change in volume during formation from the elements, ΔV, %	Crystal system	Space group	Crystal lattice		
							a	b	c
107	$MnSi_2$	113,11	21,58	(γ-Mn) −29,6	Tetragonal	—	5,51	(17,42)	3,16
108	Re_3Si	587,92	—	—	—	—	—	—	—
109	$ReSi$	214,40	16,39	−19,5	Cubic	T^4	4,775	—	—
110	$ReSi_2$	242,49	22,37	−29,8	Tetragonal	D_{4h}^{17}	3,12	—	7,65
111	Fe_3Si	195,61	27,06	−18,1	Cubic	O_h^5	5,64	—	—
112	Fe_5Si_3	363,47	55,49	−21,2	Hexagonal	D_{6h}^3	6,73	—	4,70
113	$FeSi$	83,93	13,18	−29,5	Cubic	T^4	4,438	—	—
114	$FeSi_2$	112,02	22,45	−25,4	Tetragonal	—	2,68	—	5,18
115	Co_3Si	204,91	—	—	—	—	—	—	—
116	Co_2Si	145,97	19.65	(β-Co) −21,2	Rhombic	D_{2h}^{16}	7,095	4,908	3,730
117	Co_5Si_3?	378,97	—	—	—	—	—	—	—
118	$CoSi$	87,03	13,19	(β-Co) −27,7	Cubic	T^4	4,438	—	—
119	$CoSi_2$	115,12	23,02	(β-Co) −22,6	»	O_h^5	5.35	—	—
120	$CoSi_3$	143,21	—		»	—	—	—	—
121	Ni_3Si	204,16	25,84	(β-Ni) −17,4	»	O_h^1	3,50	—	—
122	Ni_5Si_2	349,63	49,73	(β-Ni) −11,2	Hexagonal	—	7,67	—	9,75
123	δ-Ni_2Si	145,47	19,66	(β-Ni) −20,4	Rhombic	D_{2h}^{16}	7,03	4,99	3,72
	θ-Ni_2Si	145,47	18,44	(β-Ni) −25,4	Hexagonal	C_{6h}^2	3,805	—	4,890
124	Ni_3Si_2	232,25	—	—	»	—	7,63	—	9,57

constants		Micro-hardness, kg/mm^2	Specific gravity, g/cm^3	Melting point, °C	Silicon content by weight, %	Remarks
Angles	z					
—	~16	—	5,24	(Incongruent) 1144	49,67	—
—	—	—	—	—	4,79	—
—	4	—	13,08	—	13,10	$\Delta H_{form.}=-10,2$ kcal/mole
—	2	—	10,84	—	23,17	$\Delta H_{form.}=-16,6$ kcal/mole
—	4	—	7,23	—	14,36	$\Delta H_{form.}=-20$ kcal/mole
—	2	6.5 on the Mohs scale	6,55	(Stable over the range 800—1030)	23,18	$\Delta H_{form.}=-61,7$ kcal/mole
—	4	4.5 on the Mohs scale	6,37	1410	33,47	$\Delta H_{form.}=-19,2$ kcal/mole $S=12,0$ cal/deg·mole
—	1		4,99	1220	50,15	
—	—	—	—	(Incongruent) 1210	13,71	Decomposes at temperatures below 1160°
—	4	—	7,43	1332	19,24	Two modifications were reported with a conversion point at 1208°; $\Delta H°=-27.6$ kcal/mole
—	—	—	—	—	22,24	Was reported as Co_2Si_2 (24.11% Si); formed in the solid phase ≤1215° $\Delta H°=-24,0$ kcal/mole
—	4	600	6,60	1420	32,28	
—	4	580	5,00	(Incongruent) 1277	48,80	$\Delta H°=-24,6$ kcal/mole
—	—	—	—	1306	58,84	$\Delta H°=-25,6$ kcal/mole
—	1	—	7,90	Below 1040	13,76	β'-Phase of the Ni—Si system; $\Delta H°=-35.5$ kcal/mole
—	6	—	7,03	—	16,07	γ-Phase
—	4	450	7,40	(Conversion into θ) 1214	19,31	δ-Phase; $\Delta H°=-33,6$ kcal/mole
—	2	—	7,89	1290	19,31	θ-Phase
—	4?	—	—	(Decomposes in the solid phase) 830	24,19	This is apparently Ni_2Si_3, molecular weight 377.72; 22.31% ε-phase

Table 2 (continued)

Sample No.	Chemical formula	Molecular weight	Molecular volume	Change in volume during formation from the elements, ΔV, %	Crystal system	Space group	Crystal lattice		
							a	b	c
125	NiSi	86,78	14,49	(β-Ni) —20,0	Rhombic	D_{2h}^7	5,62	5,18	3,34
126	α-NiSi$_2$	114,87	23,44	(β-Ni) —20,9	Trigonal	—	8,88	—	—
	β-NiSi$_2$	114,87	23,49	(β-Ni) —20,7	Cubic	O_h^5	5,395	—	—
127	Ru$_3$Si$_2$	361,28	—	—	—	—	—	—	—
128	RuSi	129,79	15,64	—20,5	Cubic	T^4 ?	4,70	—	—
129	» Ru$_2$Si$_3$	129,79 287,67	16,58 40,92	—15,7 —19,5	» Tetragonal	O —	2,90 5,52	— —	— 4,46
130 131	Rh$_3$Si$_2$ RhSi	364,91 131,00	— 15,39	— —22,2	— Cubic	— T^4	— 4,675	— —	— —
132 133 134	Rh$_2$Si$_3$ RhSi$_2$ Pd$_3$Si	290,09 159,09 348,19	— — —	— — —	— — —	— — —	— — —	— — —	— — —
135	Pd$_2$Si	241,49	24,95	—14,7	Hexagonal	D_3^2	6,48	—	3,42
136	PdSi	134,79	17,37	—14,8	Rhombic	D_{2h}^{16}	6,121	5,588	3,374
137	Os$_2$Si$_3$	464,67	41,79	—18,5	Tetragonal	—	5,57	—	4,47
138 139	OsSi$_2$ Ir$_3$Si$_2$	246,38 635,48	— 41,75	— —14,1	— Hexagonal	— —	— 3,96	— —	— 5,12
140	IrSi	221,19	16,88	—15,8	Rhombic	D_{2h}^{16}	5,56	3,21	6,27
141 142 143	Ir$_2$Si$_3$ IrSi$_3$ Pt$_5$Si$_2$	470,47 277,37 1032,33	— — —	— — —	— — —	— — —	— — —	— — —	— — —
144	α-Pt$_2$Si	418,55	30,33	+ 2,1	Tetragonal	—	2,77*	—	2,95*
	β-Pt$_2$Si	418,55	—	—	—	—	—	—	—
145	PtSi	223,32	17,91	—13,1	Rhombic	D_{2h}^{16}	5,920	5,584	3,596
146	LaSi$_2$	195,10	37,74	(β-La) —17,1	Tetragonal	D_{4h}^{19}	4,27	—	13,74

constants		Micro-hardness, kg/mm²	Specific gravity, g/cm³	Melting point, °C	Silicon content by weight, %	Remarks
Angles	z					
—	4	400	5,99	1000	32,37	τ_1-Phase. $\Delta H°=$ $=-20,6$ kcal/mole
$\gamma =$ 90,4°	18	—	4,90	(Conversion into β) 950	48,91	ζ-Phase
—	4	—	4,89	(Incongruent) 1025	18,91	ξ-Phase
—	—	— 9.5 on the Mohs scale	—	—	15,55	—
—	4		8,30	—	21,64	—
—	1	—	7,83	—	21,64	Type CsCl
—	(2)	—	(7,03)	—	29,29	—
—	—	—	—	—	15,40	—
—	4	—	8,51	—	21,14	—
—	—	—	—	—	29,05	—
—	—	—	—	—	35,31	—
—	—	—	—	—	8,07	—
—	3	—	9,68	1250	11,63	Type Fe₂P
—	4	—	7,76	1100	20,84	Type MnP (B—31)
—	2	—	11,12	—	18,14	—
—	—	—	—	—	22,80	—
—	(1)	—	15,22	—	8,84	—
—	4	—	13,10	—	12,70	—
—	—	—	—	—	17,91	—
—	—	—	—	—	30,38	—
—	—	—	—	—	5,44	Perhaps this is Pt₅Si (4.58% Si)?
—		(Very hard)	13,8	(Changes into β) 700	6,71	Close to Fe₃B. *Data on intralattice cell
—		—	—	1100	6,71	—
—	4		12,47	1230	15,58	Pyknometric specific gravity 11.63
—	4	—	5,17	—	28,80	ThSi₂ type, as are the following lanthanide disilicides

Table 2 (continued)

Sample No.	Chemical formula	Molecular weight	Molecular volume	Change in volume during formation from the elements, ΔV, %	Crystal system	Space group	Crystal lattice		
							a	b	c
147	CeSi	168,22	29,67?	(β-Ce) — 7,6?	—	—	—	—	—
148	CeSi$_2$	196,31	36,29	(β-Ce) —16,8	Tetragonal »	D_{4h}^{19}	4,15	—	13,87
149	PrSi$_2$	197,10	34,88	(β-Pr) —20,2	»	D_{4h}^{19}	4,14	—	13,52
150	NdSi$_2$	200,45	34,21	—21,4	»	D_{4h}^{19}	4,11	—	13,45
151	SmSi$_2$	206,61	32,95	—22,4	»	D_{4h}^{19}	4,05	—	13,34
152	Th$_3$Si$_2$	752,54	76,79	— 7,1	»	D_{4h}^{5}	7,835	—	4,154
153	ThSi	260,21	28,82	— 8,2	Rhombic	D_{2h}^{16}	5,89	7,88	4,15
154	α-ThSi$_2$	288,30	37,01	—13,7	Tetragonal	D_{4h}^{19}	4,126	—	14,346
	β-ThSi$_2$	288,30	—	—	Hexagonal	D_{6h}^{1}	3,985	—	4,220
155	U$_3$Si	742,30	47,64	(γ-U) — 3,9	Tetragonal	D_{4h}^{18}	6,029	—	8,696
156	U$_3$Si$_2$	770,39	63,15	(γ-U) +4,0	»	D_{4h}^{5}	7,3297	—	3,9005
157	USi	266,16	25,59	(γ-U) +6,3	Rhombic	D_{2h}^{16}	5,66	7,67	3,91
158	β-USi$_2$	294,25	(31,81)	(γ-U) —10,6	Hexagonal	D_{6h}^{1}	3,86	—	4,07
159	α-USi$_2$	294,25	32,77	(γ-U) —7,9	Tetragonal	D_{4h}^{19}	3,98	—	13,74
160	USi$_3$	322,34	39,75	(γ-U)· —15,6	Cubic	O_{h}^{1}	4,04	-	—
161	NpSi	265,09	—	—	—	—	—	—	—
162	NpSi$_2$	293,18	32,47	(β-Np) —7,9	Tetragonal	D_{4h}^{19}	3,96	—	13,67
163	α-PuSi$_2$	295,38	32,39	(δ-Pu) —13,9	»	D_{4h}^{19}	3,97	—	13,55
164	β-PuSi$_2$	(295,38)	(32,18)	(δ-Pu) (—14,4)	Hexagonal	D_{6h}^{1}	3,884	—	4,082

constants		Micro-hardness, kg/mm^2	Specific gravity, g/cm^3	Melting point, °C	Silicon content by weight, %	Remarks
Angles	z					
—	—	—	5,67?	1530?	16,70	The specific gravity could hardly have been determined for pure CeSi (A.B.)
—	4	—	5,41	1420?	28,62	—
—	4	—	5,65	—	28,50	—
—	4	—	5,86	—	28,03	—
—	4	—	6,27	—	27,19	—
—	2	—	9,80	1700	7,47	—
—	4	—	9,03	—	10,80	—
—	4	1120	7,79	> 1670	19,49	—
—	1	—	—	8,23	—	The composition is closer to $ThSi_{1.5}$
—	4	—	15,58	(Decomposes in the solid phase) 930	3,78	—
—	2	—	12,20	1665	7,29	Was previously considered the same as U_5Si_3 (6.61% Si)
—	4	—	10,40	(Incongruent) 1575	10,55	Type FeB
—	1	—	9,25	(Incongruent) 1610	(From ~15 to 19)	Was previously considered the same as U_3Si_2 (15.04% Si) $\beta-\alpha$ conversion ? ($\Delta V = +4,2\%$)
—	4	—	8,98	1700	19,05	—
—	1	—	8,11	(Incongruent) 1510	26,14	—
—	—	—	—	—	10,60	—
—	4	—	9,03	—	19,16	—
—	4	—	9,12	—	19,02	$\Delta H_{form.} \approx -211$ kcal/mole
—	—	—	9,18	—	19,02	The composition sooner corresponds to the formula Pu_2Si_3 (14.98% Si)

BINARY SYSTEMS WITH SILICON AND THE PROPERTIES AND USE OF THE PHASES FORMED IN THEM

H—Si *System*

In 1857 Wöhler [53] first prepared monosilane, SiH_4, by treating an aluminum–silicon alloy with hydrochloric acid, but the properties of this compound and of the other saturated silicon hydrides up to Si_4H_{10} were first investigated thoroughly in 1916 by Stock [54], who proposed the name "silanes" for them.

At present, polysilanes up to $Si_{14}H_{30}$ are known, but individual compounds have been investigated only up to Si_4H_{10} (Table 2). The only fact known about the polysilanes Si_5H_{12} and Si_6H_{14} is that they are very unstable, volatile liquids [5]. Polysilanes with a greater number of silicon atoms have not as yet been isolated in a pure form.

The Si–Si chains in silanes are unstable (cf. the energies of the Si–Si and C–C bonds given above). The energy of the silicon–hydrogen bond is less than that of the carbon–hydrogen bond.

Bond	Bond energy, kcal/mole
Si–H	75.0
C–H	85.6

These phenomena explain the high chemical activity of silanes.

Monosilane is readily crystallized, but it has been studied very little in the solid state due to the very low crystallization point (88.48°K). By studying the rotation of groups forming the crystal lattice of monosilane, Clusius [55] found that the crystals formed below the solidification point had weak birefringence, while the larger crystals were completely isotropic. Monosilane apparently crystallizes similarly to methane in a cubic system. Its crystals are often twinned. However, at 63.4°K (−209.8°) there occurred an enantiotropic conversion into another modification, which had strong birefringence, and this existed to 20°K (the minimum temperature of microscopic observation in Clusius' experiments). The presence of a modification change at the given temperature was also confirmed by the course of the heat capacity against temperature curve (Fig. 2). The heat of this conversion was found to be 0.1471 kcal/mole,

which is almost of the same order as the heat of fusion of monosilane (0.1595 kcal/mole). Measurement of the heat capacity of
monosilane down to 11.35°K showed
that there were no other modification changes in monosilane at
temperatures below 63.48°K.

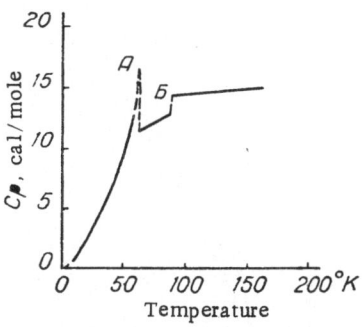

We should note that methane
has been found to show a modification change, but at a much lower
temperature (10.5°K) and with a
molar heat of conversion a factor
of almost ten less than that of monosilane.

Clusius ascribed the existence
of the modification change at 63.48°K
to the beginning of rotation of groups
present in the monosilane crystal
lattice, similar to that occurring in
some ammonium salts.

Fig. 2. Changes in the heat capacity of monosilane, SiH_4, with
temperature, according to Clusius
(1933). A) Conversion related to
the appearance of the rotation of
groups, B) melting.

The crystal lattice of monosilane is less stable than that of
methane as the data given in Table 3 show. Monosilane is readily
dissociated at temperatures above 400°.

Table 3

Thermodynamic Characteristics of Monosilane and Methane

Substance	$-\Delta H°_{25°}$ (gas), kcal/mole	Melting point, °K	Heat of fusion	
			kcal/mole	kcal/g
SiH_4	14.8	88.5	0.160	0.005
CH_4	17.9	90.7	0.224	0.014

The structures of crystalline polysilanes and unsaturated silicon hydrides have not been investigated. For Si_2H_6 with tetrahedral
angles of the Si–H bond equal to 109° 27', the Si–Si distance was
found to be 2.32 kX and that of Si–H, 1.47 kX [56] (in monosilane
the Si–H distance was found to be 1.54 kX).

The polymers $(SiH_2)_x$ and $(SiH)_x$ are solids. Their properties
have hardly been studied. The compound $(SiH_2)_x$ should presumably
contain chains as it is the hydrolysis product of CaSi [114].

Schott and Herrmann [239] obtained the polymer $(SiH)_x$ by treating $SiHBr_3$ with magnesium in ether solution. The polymer $(SiH)_x$ is a solid, yellow, friable material. It is oxidized slightly in dry air at normal temperatures and rapidly in moist air. Pure water has hardly any effect on it, but water containing alkali ions rapidly decomposes it with the evolution of hydrogen.

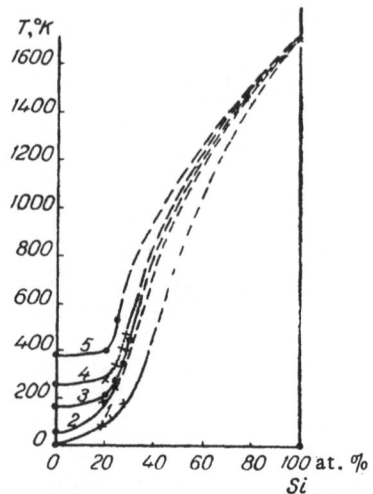

Fig. 3. Melting points of the systems: 1) H—Si; 2) F—Si; 3) Cl—Si; 4) Br—Si; 5) I—Si.

The relation of melting point to composition for the compounds investigated in the H—Si system is shown in Fig. 3.

There is no reason to consider that refractory compounds (silicon hydrides) exist in this system.

In the absence of air, SiH_4 decomposes to silicon and hydrogen at 400° and Si_2H_6 decomposes even at 250°. Other polysilanes are even more unstable and decompose in light at 0°. Solid polysilenes are converted to silanes and silicon in air [47].

Silanes ignite and explode in air. Water containing traces of alkali (for example, water in an ordinary glass vessel) decomposes silanes. In the absence of alkali, monosilane does not react with water, while Si_2H_6 reacts slowly according to the reaction equation

$$Si_2H_6 + 4H_2O = 2SiO_2 + 7H_2.$$

Dry alkalies hardly react with silanes. Likewise, acids without catalysts do not react with silanes. Alkalies decompose them with the formation of silicates by reactions of the type

$$SiH_4 + 2NaOH + H_2O = Na_2SiO_3 + 4H_2.$$

Silanes react with halogens with an explosion to give substitution products.

All silanes, and in particular polysilanes $(Si_3H_8, Si_4H_{10},$ etc.), are poisonous [285].

Silanes are used as strong reducing agents and for the synthesis of organosilicon compounds.

Halogen–Silicon Systems

It is convenient to examine compounds of silicon with halogens as substitution products of silanes. They are therefore described in this section, directly after the silanes. Figure 4 shows that the heat of formation of silicon tetrahalides falls sharply from SiF_4 to SiI_4.

Spitzer, Howell, and Schomaker [57] showed that interatomic distances in halogen compounds of a series of elements of Groups IV, V, and VI, including silicon, are in general less than the sum of the covalent radii. This indicates the existence of a partially ionic bond. Ormont [56] gave some data on this problem. Table 4 gives the characteristics of the interatomic distances of silicon halides of the type SiX_4, provided by Spitzer [57] and Dyatkina [58].

The data in Table 4 show that the ionic character of the bond decreases from SiF_4 to SiI_4, but even in the latter compound it is still partially retained. Tables 3 and 4 show that the heats of formation of all the silicon tetrahalides are considerably greater than that of monosilane. This indicates the higher stability of the former in comparison

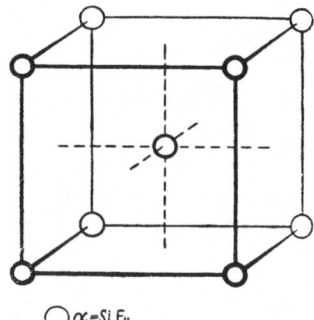

$\bigcirc \alpha\text{-}SiF_4$

Fig. 4. Crystal lattice of tetrafluorosilane, SiF_4; space group $T_d^3 - J\bar{4}3m$, $a = 5.41$ kX, $z = 2$. The Si–F distance is 1.60 kX, F–F –2.60 kX; the circles denote SiF_4 groups.

with silane. The melting point curves in the halogen–silicon systems rise more steeply in going from fluorine to iodine (see Fig. 3). The phase diagrams for these systems have not been established as yet.

F–Si SYSTEM. The properties of two compounds in the F–Si system, SiF_4 and Si_2F_6 (see Table 2) have been studied though there should be many more.

The heat of formation of silicon tetrafluoride, SiF_4 (Table 4), is greater than that of silica (209 kcal/mole) so that this fluoride may be prepared from silica. This fact is used in analytical chemistry for the quantitative determination of silica.

Mendeleev (1871) proposed the preparation of silicon tetrafluoride by heating a mixture of quartz sand, sulfuric acid, and fluorspar, using the reaction

$$SiO_2 + 2H_2SO_4 + 2CaF_2 = 2CaSO_4 + 2H_2O + SiF_4.$$

Here one should take into account the reversibility of the reaction

$$SiF_4 + 2H_2O \rightleftarrows 4HF + SiO_2,$$

and also the formation of H_2SiF_4

$$SiF_4 + 2HF \rightleftarrows H_2SiF_6.$$

Consequently, the conditions for the formation of silicon tetrafluoride should be such as to avoid if possible the decomposition of the substance.

Table 4

Some Properties of Silicon Halides of the Type SiX_4

Substance	Heat of forma- tion, ΔH°_{25}, kcal/mole	Interatomic distance, A		Compres- sion, %	Value for silicon atom after subtracting halogen radius, A
		calculated for a covalent bond	found experi- mentally		
SiF_4	370 (gas)	1.81	1.54	14.9	0.82
$SiCl_4$	145.7 (gas)	2.16	2.00	7.4	1.01
	153.0 (liquid)				
$SiBr_4$	95.1 (liquid)	2.31	2.145	7.1	1.02
SiI_4	31.6 (solid)	2.54	2.46	2.8	1.10

The thermal decomposition of barium silicofluoride may be used for the preparation of very pure silicon tetrafluoride.

$$BaSiF_6 = BaF_2 + SiF_4.$$

Silicon tetrafluoride has been known since the Eighteenth Century and was the first silicon compound synthesized that did not contain oxygen [15]. Under normal conditions, silicon tetrafluoride is a gas with a sharp smell which fumes in air. When cooled sufficiently, it is converted directly into the solid state. A liquid phase can only be obtained from silicon tetrafluoride at high pressure.

Solid silicon tetrafluoride has a body-centered cubic lattice (Fig. 4) of the Dl_2 type, consisting of SiF_4 molecules with a tetrahedral form [59]. The position of the fluorine atoms has not yet been determined accurately. The crystal lattice parameters of this compound are given in Table 2. Its thermodynamic properties were studied by Ryss [286].

Silicon tetrafluoride is used for preparing fluorine-containing organosilicon compounds and fluosilicates [663].

Hexafluorodisilane, Si_2F_6, is prepared by treating Si_2Cl_6 with fluorine. This compound has been studied very little. Its melting point ($-18.8°$) is higher than that of silicon tetrafluoride.

Cl–Si SYSTEM. Silicon tetrachloride, $SiCl_4$, may be prepared by various methods [287]. For example, carborundum, SiC (at 1000–1200°), silicon, or a mixture of silica and carbon may be chlorinated.

$$SiO_2 + 2C + 2Cl_2 = SiCl_4 + 2CO.$$

Budnikov and Shilov [60] proposed using the reaction of sulfur monochloride with silica at 1000° or passing phosgene over a mixture of silica and carbon. The processes for preparing silicon chlorides were investigated by Martin [61] and Schwarz and Pietsch [62].

The most valuable method for the preparation of silicon tetrachloride involves heating ferrosilicon (not less than 35 and preferably 50% Si) with chlorine. This process was studied in detail by Andrianov [5, 63]. Andrianov recommended the preparation of this chloride from pieces of ferrosilicon, 10–20 mm in size, loaded into a heated tube. It should be noted that alumina tubes could not be used as silicon tetrachloride reacts vigorously with Al_2O_3.

$$3SiCl_4 + 2Al_2O_3 = 3SiO_2 + 4AlCl_3.$$

The ferrosilicon was heated to 200° and then dry chlorine passed through it. The temperature rose to 450–600° due to the heat of reaction and was controlled by regulating the chlorine input. The optimal yield of silicon tetrachloride was obtained at slightly over 500° (Fig. 5). A chlorination temperature of 300–350° gave, in addition to the substance required, a small amount of polychlorosilanes (Si_2Cl_6, Si_3Cl_8, etc.), which were also obtained in the initial stages of the chlorination.

At room temperature, silicon tetrachloride is a colorless liquid, which fumes in air and has a suffocating smell; its specific gravity is 1.49. About 1% of chlorine dissolves in it and this must be removed in syntheses. Water decomposes silicon tetrachloride.

$$SiCl_4 + 2H_2O = SiO_2 + 4HCl.$$

Its chemical properties were studied by Kreshkov and Anisimov [288].

Silicon tetrachloride is very valuable in the synthesis of organo-silicon compounds and for the production of silicide films on transition metals of Groups IV, V, and VI. Silicon tetrachloride is extremely poisonous [289, 290].

Hexachlorodisilane, Si_2Cl_6, may be prepared by fractional distillation of the low-temperature chlorination products of ferrosilicon as well as by the reaction

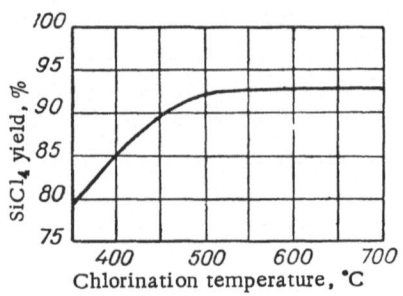

Fig. 5. $SiCl_4$ yield in relation to chlorination temperature, according to Andrianov (1955).

$$Si_2I_6 + 3HgCl_2 = Si_2Cl_6 + 3HgI_2.$$

When heated to temperatures in the range $350-1000°$, Si_2Cl_6 decomposes into silicon tetrachloride, chlorine, and silicon.

The higher polychlorosilanes may also be prepared by treating $SiHCl_3$ with a silent discharge. This gives polychlorosilanes up to Si_6Cl_{14}. These compounds are liquids at room temperature up to Si_5Cl_{12} and above this compound they are viscous, plastic substances.

Silicon tetrachloride in a stream of argon or hydrogen reacts at high temperatures with silicon to give polychlorosilanes up to $Si_{10}Cl_{22}$ [47, 64]. The latter compound is one of the inorganic substances with a very long chain of atoms. This substance is extremely reactive. When heated in vacuum, it decomposes.

$$Si_{10}Cl_{22} = 4SiCl_4 + Si_2Cl_6 + 4Si.$$

At room temperature, $Si_{10}Cl_{22}$ is a colorless, oily, viscous substance. Polychlorosilanes were studied thoroughly by Schwarz [231, 232, 233].

By heating silicon chloride in a stream of nitrogen at a pressure of 18 atm and a temperature of about 1270°, Schwarz [74] obtained an ether-soluble precipitate in the condenser tube. Removal of the ether from this solution in vacuum gave a solid, resinous, colorless film. Chemical analysis of this gave the formula $Si_{25}Cl_{52}$. However, it was not established whether this was a definite chemical compound or a mixture of polymers. Schwarz considered that this composition was the lower limit of the number of units in the chain. This experiment established that silicon compounds with very long chains could exist.

Schwarz also showed [75] that thermal decomposition (at 260–300°) of saturated higher silicon chlorides formed a polymer with the composition $(SiCl)_x$, containing 44.2 wt.% of Si theoretically, for example, by the schemes:

$$Si_{10}Cl_{20}H_2 \rightarrow SiCl + Si_3Cl_8 + Si_3Cl_7H + HCl,$$

$$Si_{10}Cl_{20}H_2 \rightarrow SiCl + SiHCl_3 + Si_4Cl_{10}.$$

Silicon subchloride $(SiCl)_x$ is a yellow, amorphous, plastic substance. It is insoluble in the usual organic solvents. It ignites in oxygen at 98° and is decomposed by water with the evolution of hydrogen. When heated, silicon subchloride decomposes by schemes of the type:

$$4SiCl = 2Si + SiCl_4,$$

$$6SiCl = 4Si + Si_2Cl_6.$$

Rochow and Didtschenko [65] established that with heating, silicon reduces silicon tetrachloride to lower chlorides. This process was investigated thoroughly by Schafer and Nickl [234].

By measuring the vapor pressure and determining the composition of the vapor, it was established that the following reaction occurred:

$$Si_{solid} + SiCl_{4\,gas} = 2SiCl_{2\,gas}.$$

The heat of formation of $SiCl_2$ is 29.9 kcal/mole and the entropy 71.1 entropy units.

This reaction is noticeable even at 800° and it proceeds with increasing vigor as the temperature is raised. The equilibrium constant

$$K_2 = \frac{P^2_{SiCl_2}}{P_{SiCl_4}} \quad mm\ Hg$$

depends on the absolute temperature T according to the equation

$$lgK_2 = 12.962 - \frac{16\,427}{T}.$$

Chlorine is also present in small amounts in the gas phase at 1327°.

It is considered that high-molecular silicon chlorides may be formed by the scheme [235]

$$SiCl_4 + (n-1)SiCl_2 = Si_nCl_{2n+2}.$$

Antipin and Sergeev [238] determined the change in the free energy of $SiCl_2$ formation from silicon and silicon tetrachloride. From their data it follows that for the latter

$$\Delta F_T^\circ = 352\ 760 - 69.27\ T,$$

$$\Delta F_T^\circ = 0 \quad \text{when} \quad T = 1527°K (\approx 1254°).$$

Therefore, with an increase in temperature the stability of $SiCl_2$ increases.

Figure 3 shows that the melting points of compounds in the Cl–Si system rise as the silicon content increases to an even greater extent than in the H–Si or F–Si systems.

Br–Si SYSTEM. Silicon tetrabromide, $SiBr_4$, is prepared by reacting bromine and silicon at red heat. This compound may also be synthesized by heating bromine with a mixture of silica and carbon. At room temperature, silicon tetrabromide is a colorless liquid, which fumes in air and is decomposed by water similarly to silicon tetrachloride. So far, silicon tetrabromide has no special technical value. Polybromosilanes up to Si_4Br_{10} are prepared by the action of a silent discharge on $SiHBr_3$ or by reacting bromine with calcium silicides.

Hexabromodisilane may also be prepared by the exchange reaction

$$Si_2I_6 + 3Br_2 = Si_2Br_6 + 3I_2.$$

Oxidation of hexabromodisilane forms silica [66].

Reduction with silica at 1200° gives a polymer, $(SiBr_2)_x$, which is a brown, friable, tarlike substance (14.95% Si); this substance dissolves in ligroin and other nonpolar liquids. It is converted into SiO_2 above 100° in air. When heated in the absence of air above 200°, $(SiBr_2)_x$ is cleaved to give Si_2Br_6 and silicon subbromides. At 350° it is converted into the polymer $(SiBr)_x$ (26.01% Si), and at 550–600° it gives silicon. Determination of the molecular weight gave a value of 3000, which corresponds to $x = 16$.

The compound $(SiBr_2)_x$ gives various organosilicon derivatives. It is formed by heating Si_2Br_6 to 300° for 6–8 hours with the simultaneous removal of $SiBr_4$ [708].

Figure 3 shows that with an increase in the silicon content of bromosilanes, the melting points rise by the same law as for chlorosilanes.

Compounds of silicon with bromine have acquired no practical value as yet.

I–Si SYSTEM. Silicon tetraiodide, SiI_4, is prepared by passing a mixture of iodine vapor and CO_2 over silicon at a temperature of about 500°. This compound is a white, crystalline, very hygroscopic substance. When in the vapor state, silicon tetraiodide ignites spontaneously in air. Table 4 shows that it must be considerably less stable than other silicon tetrahalides and this is actually

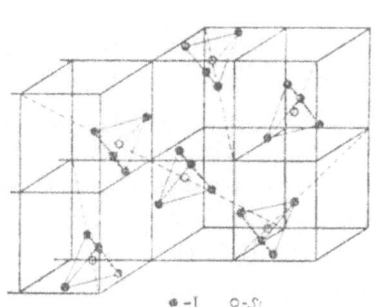

Fig. 6. Crystal lattice of silicon tetraiodide, SiI_4, according to Hassel and Kringstad (1931).

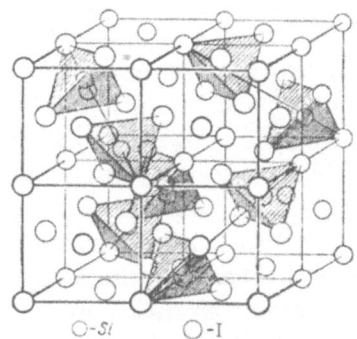

Fig. 7. Crystal lattice of silicon tetraiodide, SiI_4, according to Ormont. DI_1, type, space group $T_h^6 - Pa\,3$, $a = 11.9$ A, $z = 8$.

the case. Silicon tetraiodide crystallizes in a cubic system [67] of the same type as GeI_4 and SnI_4 [68]. Its crystal lattice consists of tetrahedral groups (Fig. 6) in which the silicon atom is surrounded by four iodine atoms. The center of gravity of each such tetrahedron has one neighbor at a distance of $\frac{1}{4}$ of the solid diagonal of the cube and three neighbors approximately 25% further away. The crystal lattice of silicon tetraiodide is shown in Fig. 7 in a somewhat simplified form. It shows clearly the island-type structure.

Hexaiododisilane, Si_2I_6, was prepared by Friedel in 1868 by heating silicon tetraiodide to 300° with finely powdered silver.

$$2SiI_4 + 2Ag = Si_2I_6 + 2Ag.$$

This compound consists of birefringent, six-faced prismatic crystals, which sublimed in vacuum at 250° with partial decomposition. The compound melts at 250° [69].

The compound $(SiI)_x$ was prepared by thermal decomposition of Si_2I_6 [703].

It follows from Fig. 3 that silicon iodides have the highest melting points of all the silicon halides. The melting points of silicon iodides increase sharply with an increase in the silicon content.

So far the practical value of silicon iodides is limited to their use in syntheses of organosilicon compounds.

At–Si SYSTEM. The At–Si system has not been studied at all and no compounds of astatine and silicon have been prepared.

Systems Formed by Silicon with Elements of Group IA

With the exception of Li_6Si_2, silicides of alkali metals were unknown until recently and it was even considered that they did not exist at all. However, the silicide $Na(NaSi_2)$ was prepared in 1947 and then a series of other sodium, potassium, rubidium, and cesium silicides, although the phase diagrams of silicon–alkali metal systems and the thermodynamic and other constants of the compounds formed in these systems have not been established up to now.

Silicides of alkali metals have been studied very little and, in particular, their melting points are still unknown. We may surmise that the melting points of these compounds are low. Alkali metal silicides have no practical use as yet.

Li–Si SYSTEM. According to Moissan [70], sintering of silicon with excess lithium in vacuum at red heat gives lithium silicide. The silicide is freed from the metal by extraction with liquid ammonia, which dissolves lithium, and, preferably, by distillation of the lithium in vacuum at 400–500°. The silicide decomposes above 600° with volatilization of the lithium. From analysis results, Moissan ascribed the formula Li_6Si_2 (57, 14% Si), to this silicide and it is still used in the literature. However, Klemm and Struck [71] showed that this compound does not exist but that there are two lithium silicides, Li_4Si and Li_2Si, which were prepared in nickel crucibles. Arvin Weiss and Alaric Weiss determined the composition of the first silicide more precisely and showed that it corresponded to the formula $Li_{15}Si_4$ (52.0% Si). This silicide is apparently isomorphous with $Na_{15}Sn_4$.

These circumstances seem to indicate that Moissan obtained $Li_{15}Si_4$ and not Li_6Si_2. The properties of lithium silicides are given in Table 2, using the data of Moissan and Klemm.

The lithium silicide obtained by Moissan was not reduced by hydrogen when heated to 600°. It ignited when heated moderately with fluorine and when heated to red heat with chlorine and bromine and iodine vapor. Sulfur, selenium, tellurium, and phosphorus decomposed lithium silicide. This compound is a strong reducing agent. Chromium, iron, manganese and calcium oxides are reduced by lithium silicide, while alumina does not react with it.

Lithium silicide reacts with water very vigorously and burns brightly in air.

The study of the Li–Si system cannot be considered complete. For example. it is not known whether there are lithium silicides that are richer in silicon than Li_2Si. similar to the silicides of other alkali metals.

Na–Si SYSTEM. Nowotny and Scheil [72] were the first to prepare sodium silicide by fusing sodium with silicoaluminum in a fire-clay crucible and it was essentially a ternary compound containing 15.3 at.% Na, 18.0 at.% Al, and 66.7 at.% Si, which corresponds to the formula $NaAlSi_4$ or $(Na, Al)Si_2$. This compound belongs to the tetragonal system ($a = 4.13$ A, $c = 7.40$ A, six atoms in the elementary cell), apparently with the space group D_{4h}^7 ($C\,38$ type). The sodium atoms are arranged in layers in the lattice. The packing here is more dense than in $MoSi_2$, for example. Pure sodium disilicide was then obtained by reducing quartz with metallic sodium. It is less stable than the ternary compound described above but its structure is similar, the only difference being that the c axis is doubled. In contrast to the ternary compound, sodium disilicide is completely soluble in water. Table 2 shows that this compound is formed with a very great increase in volume (by almost $\frac{1}{3}$), which indirectly indicates its low stability.

Hohmann [73] obtained a series of alkali metal silicides (sodium, potassium, rubidium, and cesium) by direct synthesis from the elements in corundum crucibles in an argon atmosphere. A 3–4 fold excess of alkali metal was used. Sodium silicide, NaSi, was prepared by heating the substances at 700° for 24 hours. Excess sodium was removed by vacuum distillation. If the sodium silicide obtained was poorly crystallized it ignited in air. This compound reacts with water and acids. It was found that when sodium monosilicide was heated in vacuum in a glass tube it decomposed at 240°, leaving silicon while the sodium volatilized. In this case, no sodium silicides with a high silicon content were formed as was the case in the thermal decomposition of potassium, rubidium, and cesium monosilicides.

The crystal structure of sodium silicide has not been investigated.

K–Si SYSTEM.   Hohmann [73] also prepared potassium mono-silicide, KSi, similarly to sodium monosilicide at a temperature of 650°. KSi crystallizes poorly. It is oxidized in air very rapidly with an explosion. Its other properties are similar to those of sodium monosilicide. Potassium monosilicide is decomposed at 360° in vacuum, leaving a substance which differs from the original one and from silicon according to x-ray pictures. From the weight of the residue and the calculated silicon content, Hohmann concluded that this was a new potassium silicide with the composition KSi_8. As with rubidium and cesium polysilicides, the composition of this silicide cannot be considered as established conclusively as no direct chemical analysis has been carried out. The homogeneity of these substances was not established either.

Rb–Si SYSTEM.   Hohmann [73] also prepared rubidium mono-silicide, RbSi, at a temperature of 600° in the form of fine dark crystals. A comparison of x-ray pictures showed that potassium and rubidium monosilicides were of the same type. The chemical properties of rubidium monosilicide are similar to those of KSi, though rubidium monosilicide is more active chemically than potassium monosilicide. Its decomposition point in vacuum is 350–360°. The residue after its decomposition was a silicide with the composition $RbSi_8$ and, judging by x-ray pictures, of the same type as KSi_8.

Cs–Si SYSTEM.   Cesium monosilicide, CsSi, was prepared by the same method as NaSi [73] (but at 600°) in the form of friable crystals. This substance is even more active chemically than rubidium monosilicide. The x-ray picture for cesium monosilicide indicates that this monosilicide is not of the same type as RbSi (many new weak and diffuse lines). Cesium monosilicide decomposes in vacuum even at 300–350°. The residue from its thermal decomposition in vacuum was a mass which was assumed to correspond to $CsSi_8$ as in the case of potassium.

Fr–Si SYSTEM.   This system has not been studied at all. No francium silicides have been prepared.

Systems Formed by Silicon with Elements of Group IB

Systems formed by silicon with elements of Group IB are of great practical value and have therefore been investigated extensively, especially the Cu–Si system. However, up to the present time, no exhaustive data have been obtained which would make it possible to elucidate all the structural details of the Cu–Si, Au–Si, and, to a lesser degree, Ag–Si systems. We can assume that this gap will be closed in the near future by investigations with modern procedures.

Cu–Si SYSTEM. The first investigations of the Cu–Si system were carried out in the last century by Berzelius (1824), St. Claire Deville and Coron (1857), and then continued by Vigouroux (1896), de Chalmot (1896–1897), and Lebeau (1905). As a result, the formation of copper silicides was established but their compositions were not always determined. Vigouroux [76] considered that he was able to prepare a silicide with the composition Cu_2Si by fusing copper and silicon. Lebeau [77] concluded from his preparative work

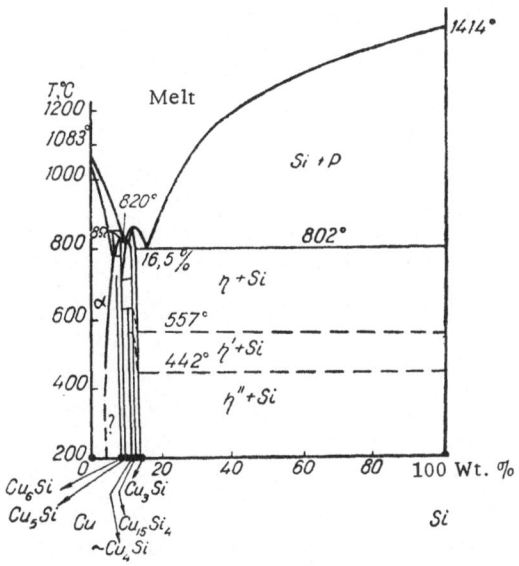

Fig. 8. Phase diagram of the Cu–Si system, according to Smith (1928) and Jokibe (1931).

that copper and silicon form the silicide Cu_4Si (9.94% Si), which he described in a series of publications (1905–1906). These investigations are now mainly of historical interest as are a number of subsequent publications describing various copper silicides that have since been proved nonexistent. As Hansen [28] has examined them in detail, there is no need to dwell on them here.

As a result of extensive investigations, Smith [78] constructed the phase diagram of the Cu–Si system and this is usually included in textbooks, even at the present time (for example, [79]), despite inaccuracies in individual details.

Jokibe [80] corrected some of Smith's data, especially those referring to the conversion temperatures of the η-phases in the solid state.

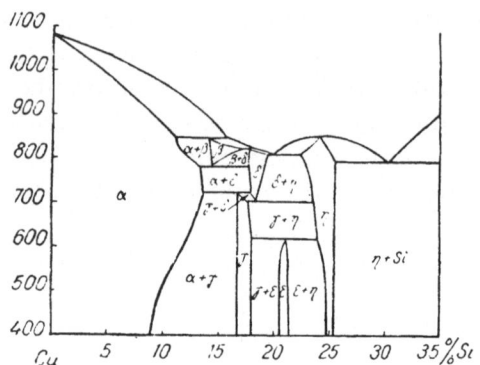

Fig. 9. A detail of the phase diagram of the Cu−Si system.

The structure of separate phases in the Cu–Si system was investigated by Arrhenius and Westgren [81], who also found a phase (ϵ-$Cu_{15}Si_4$) which Smith had not detected previously. These investigations made it possible to construct a more or less accurate phase diagram of the Cu–Si system, which is given in Figs. 8 and 9. Supplementing them, Table 5 gives data on the phase equilibria of this system.

Using a high-temperature x-ray camera, Tarasova [82] established that the solubility of silicon in copper decreases from 4.6 wt.% at 500° to 1.1% at 170°. The solubility curve of silicon in copper is not logarithmic at these temperatures. The structure of the α-phase, which is a solid solution of silicon in copper, is a face-centered cubic one. Its region of homogeneity may be seen in Figs. 8 and 9. The size of the elementary cube in the crystal lattice changes from 3.608 for pure copper to 3.615 A at the saturation limit [81].

The following β-phase of the Cu–Si system is homogeneous at approximately 7.0 wt.% of Si and is stable only at high temperatures (Table 5). Its composition corresponds to the formula Cu_6Si and the electron concentration per atom is 1.44 [83]. The properties of this phase are given in Table 2.

Table 5

Phase Equilibria in the Cu–Si System According to Smith,
Jokibe, Arrhenius, and Westgren

Temperature, °C	Characteristics of conversion	Phase present and silicon content, wt.%
852	Peritectic	α-Solid solution (5.25% Si)+ Melt (7,7% Si) ⇄ ⇄ β-Phase (6,8% Si)
824	»	β-Phase (8,4% Si) + Melt (8,9% Si) ⇄ δ-Phase (8,6% Si)
820	Eutectic	δ-Phase (9,7% Si) + η-Phase (11,2% Si) ⇄ Melt (9,9%)
802	»	γ-Phase (12,8% Si) + Si ⇄ Melt (16,0% Si)
782	Reaction	β-Phase (7,75% Si) ⇄ α-Phase (6,7% Si) + δ-Phase (8,60% Si)
710	»	δ-Phase (8,95%) ⇄ γ-Phase (8,60% Si) + η-Phase (11,7% Si)
557	Phase change	η-Phase (11,75% Si) ⇄ η′-Phase (11,75% Si)
442	"	η′-Phase (11,75% Si) ⇄ η″-Phase (11,75% Si)
800	Reaction	δ-Phase + γ-Phase ⇄ ε-Phase

The γ-phase of the Cu–Si system is a typical electronic compound with a composition corresponding to the formula Cu_5Si and a β-Mn structure (Fig. 10) with a statistic distribution of copper and silicon atoms in the crystal lattice [81, 84, 85]. The electron concentration in this phase is 1.54 [83, 86]. Its region of homogeneity is close to 8.3 wt.% of Si. The other properties of this phase are given in Table 2.

One can see from Table 5 and Figs. 8 and 9 that the δ-phase has a composition which is close to the formula Cu_4Si and is stable only at high temperatures. It is homogeneous at approximately 8.8 wt.% Si. This phase has been studied little as yet.

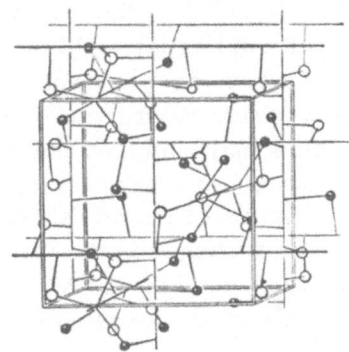

O Me$_I$, ● Me$_{II}$

Fig. 10. Crystal lattice of the β-Mn type in which Cu_5Si crystallizes.

Arrhenius and Westgren [81] found that the Cu–Si system included another phase, which they called the ε-phase and which is stable over the range from room temperature to 600° and to 800° according to Jokibe [80]. In analogy with the Cu–Sn system, this phase is sometimes [83, 86] given the composition $Cu_{31}Si_8$ (10.2% Si).

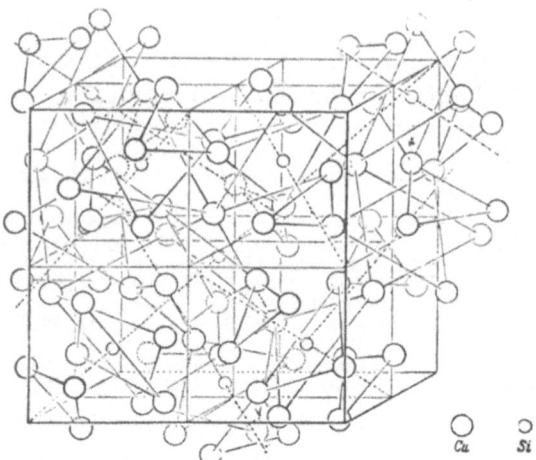

Fig. 11. Crystal lattice of $Cu_{15}Si_4$, $D8_6$ type. For the sake of clarity, some atoms are not shown.

However, an x-ray investigation of it [87] showed that its composition was $Cu_{15}Si_4$ (10.5% Si) and that it had a peculiar structure, which is shown in a simplified form in Fig. 11. A similar structure was found for the electronic compounds $Na_{15}Pb_4$ and $Na_{15}Sn_4$ by Zintl and Harder [1936].

According to Arrhenius, the narrow region of homogeneity of this phase corresponds approximately to 10.5 wt.% of Si, while others [83] indicated that this region is in the range 15.2–16.4 wt.% of Si with an electron concentration of from 1.58 to 1.62.

Figure 11 shows that the silicon atoms are surrounded by copper atoms arranged in the form of four equilateral triangles. The coordination number for the silicon atom is 12. Each silicon atom has the following closest neighbors:

Neighbor	Distance, A
$3Cu_{II}$	2.38
$3Cu_{II}$	2.48
$3Cu_{II}$	2.62
$3Cu_{I}$	2.62

The silicon atoms are at the centers of gravity of the triangles formed from three Cu_I atoms, through which the trigonal axis of symmetry passes normally to the planes of these triangles.

Additional information on the $Cu_{15}Si_4$ phase is given in Table 2.

The last binary phase of the Cu–Si system, the η-phase, was found and partially investigated by Smith [78] and Jokibe [80]. It is homogeneous at approximately 12.8 wt.% of Si and therefore its composition corresponds to the formula Cu_3Si, which is given, for example, in [86]. As in the case of the η-phase, the δ-phase has sometimes been assigned the formula Cu_4Si, but this apparently has less basis than the first formula. The electron concentration in the η-phase is 1.75 [86]. The nature of this phase and its conversions in the solid state at 557 and 442°, established by Jokibe [80], have not been investigated as yet.

Table 6

Properties of Copper and Magnesium Silicides

Formula	Crystal system	Space group	Lattice parameters			Remarks
			a	c	z	
Cu_3Mg_2Si	Hexagonal	D_{6h}^4, type C^{14}	5.00	7.87	4	Low-temperature form
C_3Mo_2	Hexagonal	D_{6h}^4, type C^{36}	5.0	16.0	8	High-temperature form
$Cu_{16}Mg_9Si_7$	Face-centered cubic	—	11.67	—	4	

All copper silicides appear like metals. They react [23] with dilute acids and, when heated strongly, with chlorine and sulfur.

It has been established that copper dissolves in silicon [140], but in extremely small amounts, which increase somewhat with an increase in temperature.

Copper also forms silicides in ternary systems. Table 6 gives some data on the binary silicides of copper and magnesium.

In the low temperature modification of Cu_3Mg_2Si, the copper and silicon atoms are arranged regularly so that a superlattice is formed, while in the high-temperature modification, they are distributed statistically. These modifications, which crystallize similarly to $MgZn_2$ and $MgNi_2$, respectively, are Laves phases.

There are indications that their structures are similar to $MgCu_2$ [83] with a composition represented as $MgCu_{1.5}Si_{0.5}$. The first treatment is presumably more accurate. The character of the bonds in them, as in the compound $Cu_{16}Mg_9Si_7$, requires further study.

The data given indicate the very complex structure of the phase diagram of the Cu–Si system.

This system is of great practical importance. Copper-silicon alloys are used instead of tin-bronze as they have suitable mechanical properties and a lower specific gravity. Such bronzes usually have a silicon content of 2-5%. Manganese is often added (BrKMr 3-1 grade, used for the manufacture of wire, strips, and rods [79]). Copper bronze with 4% Si and 2% Mn is sometimes called "Everdur." Copper silicides are used for the synthesis of other metal silicides.

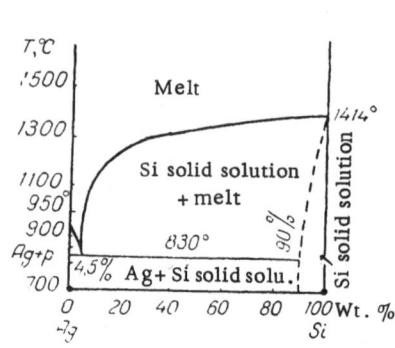

Fig. 12. Phase diagram of the Ag—Si system according to Arrivaut (1908).

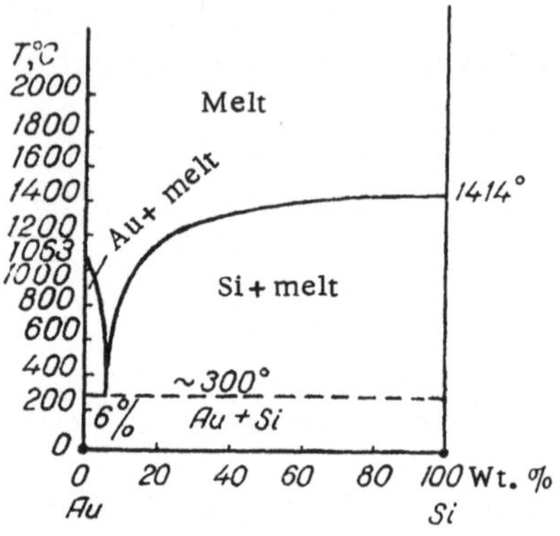

Fig. 13. Phase diagram of the Au—Si system according to Di Capua (1920), Loskiewicz (1929), and Jette and Gebert (1933).

Ag–Si SYSTEM. Contrary to the Cu–Si system, the Ag–Si system is very simple. No chemical compounds are formed in it as can be seen from the phase diagram established by Arrivaut [88]

Table 7

Crystal Lattice Parameters in Silver–Silicon Alloys [89]

Material	Annealing temperature, °C	Lattice parameters a, A	
		Ag	Si
Pure Si	—	—	5,4170
Pure Ag	—	4,0765	—
Alloy with 5% Si	825	4,0769	5.4174
» » 40% Si	635	4,0765	5.4176
» » 40% Si . . .	635	4,0768	5.4186
» » 40% Si	635	4,0771	5.4173
» » 40% Si . . .	825	7,0769	5.4181
Alloy with 50% Si	770	4,0764	5.4167
» » 70% Si	635	4,0773	5.4175
» » 80% Si	825	—	5.4176
» » 85% Si	750	4,0761	5.4174

(Fig. 12). The solubility of silicon in silver has not been investigated. The maximum concentration of silver in a solid solution with silicon was determined approximately from the results of thermal analysis and the microstructures of unannealed preparations.

Loskiewicz [28] reported that the Ag–Si system has solid solutions on the silver side as well. This, however, was not confirmed by the investigations of Jette and Gebert [89], who showed by x-ray investigations that solid solutions are hardly formed in this system, as can be seen from the data in Table 7. In any case, the solubility of silicon in silver is less than 0.8 wt. %.

The practical value of the Ag–Si system is extremely limited as yet.

Au–Si SYSTEM. Figure 13 gives the phase diagram of the Au–Si system according to Di Capua [90] with a corrected eutectic point [91] and indications of solid solutions removed [89]. It should be noted that by studying the diffusion of silicon in gold, Loskiewicz came to the conclusion that silicon was soluble in gold. However, the x-ray investigation of gold and silicon alloys (50% Si, annealing temperature 240°) carried out by Jette and Gerbert [89] showed the constancy of the lattice parameters of gold and silicon (the gold lattice parameter in the alloy equalled 4.0682 A and that of silicon, 5.4176 A). It follows from this that no solid solutions with an appreciable concentration of the second component are formed.

The Au–Si system has no practical use as yet.

Systems Formed by Silicon with Elements of Group IIA

The systems formed by silicon with elements of Group IIA differ greatly in the structure of their phase diagrams. They have also been studied to different extents. It is therefore impossible to make any general remarks.

Be–Si SYSTEM. The phase diagram of the Be–Si system according to Masing and Dohl [93] is shown in Fig. 14. To supplement this, Sloman [94] carried out a microscopic investigation of alloys rich in berryllium and reported that silicon is apparently insoluble in crystalline beryllium.

Figure 14 shows that the Be–Si system is a simple eutectic one. No beryllium silicides have been obtained up to now. Indications that they exist, occasionally found in reviews, have no foundation.

In ternary systems containing beryllium and silicon, mixed silicides may be formed and an example of one is the compound ZrBeSi

[138], which has a hexagonal structure with a lattice of the NiAs type and the parameters $a = 3.71$ A, $c = 7.19$ A, $z = 2$; its specific gravity is 4.95; the compound should presumably be considered as a derivative of $ZrBe_2$, which also has a hexagonal structure, but of a somewhat different type (AlB_2).

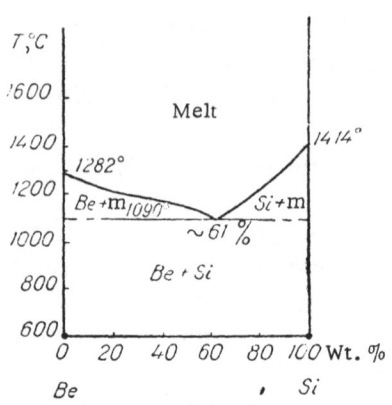

Fig. 14. Phase diagram of the Be−Si system according to Masing and Dohl (1929).

The Be−Si system has no practical value in itself at present, but is partly used in complex beryllium-containing melts.

Mg−Si SYSTEM. The existence of magnesium silicides was first discovered in the middle of the Nineteenth Century by Martius in slag when he was preparing magnesium [23]. Winkler [95] established that the magnesium silicides with the composition MgSi, Mg_4Si_3, Mg_5Si_3, and others that were described by Wöhler (1858) and Geiter (1865) did not exist, apart from Mg_2Si. Lebeau and Bossuet [96] investigated microscopically, alloys of the Mg−Si system from 0.4 to 77.2% Si and confirmed Winkler's conclusion.

The first phase diagram of the Mg−Si system was plotted by Vogel [97] and then refined by Wöhler and Schliephake [98].

The extraction of free magnesium from alloys with bromobenzene yielded no residue richer in magnesium than Mg_2Si. Therefore, most attention was paid to compositions between Mg_2Si and MgSi. Wöhler and Schliephake showed that the gray-blue silicide, Mg_2Si, formed in mixtures of magnesium and silicon at 600−700°. With heating to 800° and above, part of the magnesium was volatilized to give a mixture with a different composition, which had previously been considered as various complex magnesium silicides. The data on the binding of magnesium with silicon seemed to indicate the formation of a silicide with the composition MgSi, which is stable only at high temperatures.

When they heated samples to the melting point (1100−1200°) and cooled them gradually, Wöhler and Schliephake obtained only Mg_2Si;

only with a low magnesium content did they obtain a silicide with the probable composition of MgSi. Wöhler and Schliephake considered that when MgSi was cooled below 600° it decomposed to Mg_2Si and silicon. When a melt with the composition of MgSi was quenched in water, this silicide was not found.

In general, the phase diagrams of the Mg–Si system constructed by Vogel [97] and by Wöhler and Schliephake [98] are exactly the same. The liquidus temperature given by the latter authors is almost 30° lower than that of Vogel, though the preparations contained fewer impurities.

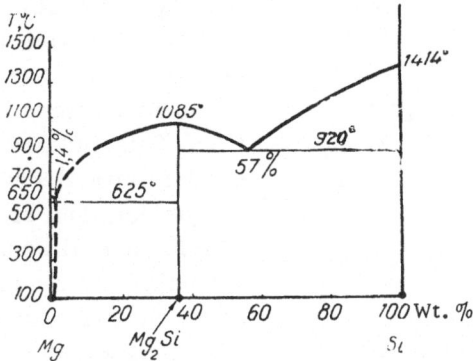

Fig. 15. Phase diagram of the Mg–Si system. The solid line is according to Wöhler and Schliephake (1926) and the dotted lines are according to Vogel (1909).

Figure 15 shows the phase diagram of the Mg–Si system, based on data of Wöhler and Schliephake with the composition of the Mg–Mg_2Si eutectic corrected by Schmidt [99]. Mannchen [100] measured the electrical conductivity of magnesium–silicon preparations and showed that magnesium dissolved a small amount of silicon, but he did not establish the limit of solubility or its temperature dependence. Thus, we may consider that at present the existence of only one silicide, Mg_2Si, has been established in the Mg–Si system. Therefore, references to other magnesium silicides, which are occasionally found in descriptions of silane preparation, require correction.

The structure of the crystal lattice of dimagnesium silicide, Mg_2Si, was investigated by Owen and Preston [101] and by Klemm and Westlinning [102]. As a result it was found that the compound has a fluorite type of structure (Fig. 16) with the parameters given in Table 2.

On the basis of x-ray investigations, Klemm and Westlinning [102] also established that the region of homogeneity of dimagnesium silicide in the Mg–Si system is very small. The solubility of silicon in solid magnesium is also very small. Wöhler and Schliephake

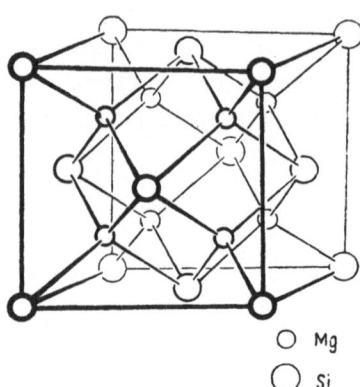

Fig. 16. Crystal lattice of Mg₂Si.
CaF₂ type, space group $O_h^5 - Fm3m$, $a = 6.338$ kX, $z = 4$.

[98] found that when this compound was heated for a long time, magnesium volatilized. Kubaschewski and Villa [106] determined its heat of formation (see Table 2).

Determination of the electron density in dimagnesium silicide crystals showed [542, 543] that it decreased to almost zero between the silicon atoms in the [100] direction. This indicates the absence of conductivity electrons in Mg₂Si. The magnesium and silicon atoms in this silicide are almost unionized, while the polar forces are very weak. The bond strength corresponds to an electron density of 0.2 e/A³. This magnesium silicide is a coordination compound. It is very friable and has a high electrical resistance. Busch and Winkler [103] established that after all, dimagnesium silicide is electrically conducting.

Savitskii and Baran [633] studied the microhardness of the Mg–Si system.

Thin sections of dimagnesium silicide have a brown-red color in transmitted light. Water decomposes it slowly and acid, vigorously. In an atmosphere of hydrogen at 1100–1200°, molten magnesium silicide decomposes with volatilization of magnesium. Heating it in a current of nitrogen gave a black mass, which evolved ammonia in a moist atmosphere, evidently due to the formation of magnesium nitride.

The Mg–Si system is very important in the technology of magnesium and aluminum alloys. Lithium–magnesium alloys with 1–1.5% Si (ML1) are used for manufacturing simple apparatus for use under hermetically sealed conditions. Only dimagnesium silicide was formed in the ternary Al–Mg–Si alloys also [104, 105]. This compound dissolves slightly in aluminum (Fig. 17). Such solid solutions may be heat treated and they are widely used in practice, especially with the addition of small amounts of manganese [79], for example, Aldrey (0.6% Mg, 0.6% Si, and 0.6% Mn), which is readily worked and has a high electroconductivity. Magnesium and silicon are also found in Duralumin. Only dimagnesium silicide is present in the Cu–Mg–Al–Si system [105]. Its presence in aluminum increases the hardness of the composition [243].

This magnesium silicide is used also for preparing silanes (see above).

Ca–Si SYSTEM. The first calcium silicide was $CaSi_2$, which was prepared by Wöhler (1863) and then by a number of other investigators. Le Chatelier [107] showed that one other calcium silicide exists. Hackspill [108] assigned the formula Ca_3Si_2 to this compound. Kolb [109] considered that calcium silicides with the composition Ca_6Si_{10} and $Ca_{11}Si_{10}$ exist. However, it was later established that the last three calcium silicides do not exist, while the silicide, CaSi does [110]. Regrettably, Kolb's archaic data are still cited in the literature.

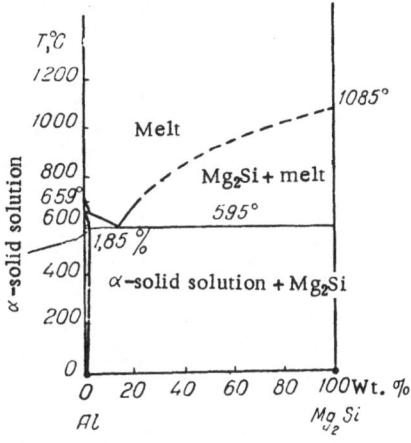

Fig. 17. Phase diagram of the pseudo-binary system $Al–Mg_2Si$.

Tamaru [111] undertook to establish the phase diagram of the Ca–Si system, but the results he obtained are unreliable due to the impurity of the silicon (92.5% in a series of experiments) and contamination of the preparations with nitrides as the former were melted in a nitrogen atmosphere. The phase diagram of the Ca–Si system was developed satisfactorily only by Wöhler and Schliephake [98], who studied alloys with an Si content from 15 to 86% and as a result another calcium silicide (Ca_2Si) was found (Fig. 18). In these investigators' experiments, 98.5% pure calcium and 99.5% pure silicon were used. The phase diagram was plotted from thermoanalysis data. Mixtures of calcium and silicon were placed in a crucible of magnesia, zirconium dioxide, or equal amounts of Al_2O_3 and MgO since alumina crucibles decomposed rapidly. Chromium dioxide was added to the molten composition of the crucibles to increase their stability. All the melts were analyzed. The phase diagram of the Ca–Si system according to Wöhler and Schliephake is given in Fig. 18, from which it follows that only the monosilicide CaSi melts congruently, while the other two (Ca_2Si and $CaSi_2$) melt incongruently. The Ca–Ca_2Si eutectic was determined solely by extrapolation as metallic calcium destroyed the crubibles very drastically so that a direct determination was impossible. The

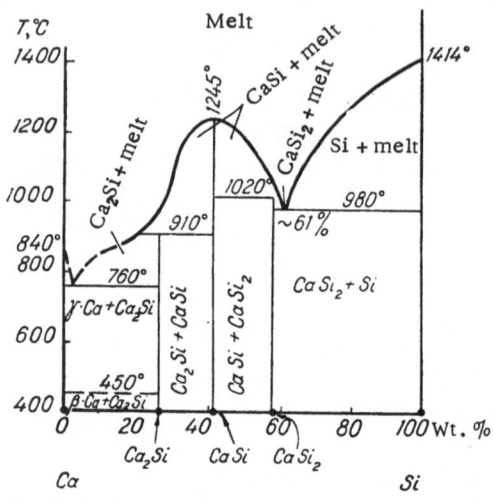

Fig. 18. Phase diagram of the Ca–Si system according to
Wöhler and Schliephake (1926).

presence of three calcium silicides in the Ca–Si system was also established from measurements of the electric potentials of the melts. The phase relationships in the Ca–Si system may be seen from the diagram in Fig. 18 and needs no explanation. The modification changes of calcium are based on the latest investigations. The effect of silicon on them has not been studied as yet. Louis and Frank [112] experimented with very pure preparations and found that, contrary to Wöhler's findings, calcium monosilicide does not decompose to calcium disilicide at very high temperatures. These authors considered that dicalcium silicide was a compound with cubic symmetry, but this was not confirmed by later investigations of Eckerlin and Wolfel [113], who were able to synthesize this silicide in a corundum crucible in an argon atmosphere at the melting point (1200°). According to these authors, dicalcium silicide has rhombic symmetry with a $PbCl_2$ type of lattice (Fig. 19) and the elementary cell parameters that are given in Table 2. There is no Si–Si bond in this silicide.

Louis and Frank [112] considered that calcium monosilicide underwent modification changes when heated. However, Helner [114] showed by high-temperature x-ray investigation (up to 1000°) that this hypothesis was incorrect: calcium monosilicide exists in only one form – a rhombic one with the structure illustrated in Fig. 20. Angular chains of silicon atoms

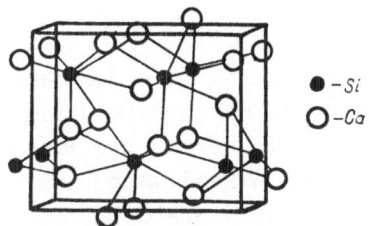

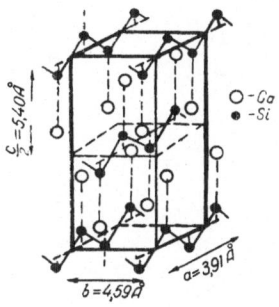

Fig. 19. Ca_2Si crystal lattice according to Eckerlin and Wölfel (1955); C23 type; space group D_{2h}^{16} — $Pnma$, $a = 9.002$ A, $b = 7.667$ A, $c = 4.799$ A, $z = 4$.

Fig. 20. CaSi crystal lattice according to Helner (1950); space group D_{2h}^{17} — $Cmmc$; $a = 3.91$ A. $b = 4.59$ A, $c = 10.795$ A, $z = 4$. Coordination: Si : 7Ca + 2Si, Ca: 7Si + 6Ca.

are disposed in the direction of the a axis, while the calcium atoms are above the apexes of the angles. The angle at the chain units is 10.5°. The distance between two adjacent silicon atoms in a chain is 2.47 A, which indicates homeopolar bonds in the chain. The Ca–Si distance is 3.11 A and this indicates that there is a weakly expressed ionic bond here. Calcium monosilicide is characterized by the absence of tetrahedral surroundings for the silicon atoms. The crystal structure of calcium disilicide was investigated by Böhm and Hassel [115]. The results of their investigations are given in Fig. 21 and in Table 2.

The calcium disilicide structure contains $[Si_6]$ rings. The Ca–Si distance equals 3.06 A, which indicates an ionic bond.

The heats of formation of calcium silicides (see Table 2) were determined by Kubaschewski and Villa [106]. The almost zero heat of the reaction

$$CaSi + Si = CaSi_2.$$

as found in their data. is quite remarkable. The reason for this is still unknown.

Dicalcium silicide, Ca_2Si, is a metallike substance, which gradually decomposes in air. This compound reacts noticeably with alcohol and even more so with water. It is decomposed by hydrochloric acid without the formation of silanes (in contrast to calcium monosilicide). When it is heated to 900° in a hydrogen atmosphere, the following reaction occurs [112]:

$$Ca_2Si + H_2 = CaSi + CaH_2.$$

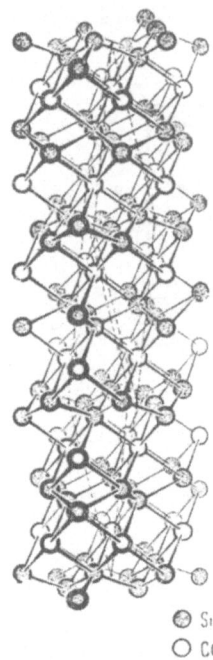

○ Si
○ Ca

Fig. 21. $CaSi_2$ crystal lattice according to Böhm and Hassel (1927); space group $D_{3d}^5 - R\bar{3}m$; $z = 2$. Co-ordination Ca : 6Si, Si : 3Ca. The Si−Si distance equals 2.48 kX, Ca−Si−2.99 kX, Ca−Ca−3.88 kX.

Calcium monosilicide, CaSi, has a metallic appearance; it is decomposed by hydrochloric acid to form silicon hydrides, including $(SiH_2)_x$ [114]. When heated in a stream of nitrogen, it forms calcium silicocyanamide.

$$CaSi + N_2 = CaSiN_2.$$

Calcium disilicide is decomposed by water and humid air to give hydrogen [23]:

$$CaSi_2 + 6H_2O = Ca(OH)_2 + 2SiO_2 + 5H_2;$$

reaction with HCl does not give silanes, but a yellow silicone [98] and possibly Si_2H_2. It reacts with fluorine with ignition. When heated in air, it is oxidized vigorously on the surface. It reacts with sulfur at red heat. When heated with carbon, it is converted to calcium carbide and carborundum.

Calcium disilicide decomposes when heated in a nitrogen atmosphere [116]. This gives calcium silicocyanide, $Ca(SiN)_2$, and silicocyanamide, $CaSiN_2$, elementary silicon and also small amounts of silicon and calcium nitrides. Calcium disilicide is rapidly converted into the monosilicide. Calcium disilicide reacts with aluminum at 1000° to give the compound $Ca_2Al_4Si_3$ [241].

Calcium silicides are also used in the synthesis of silicides and silicon hydrides.

Calcium silicides are used as reducing agents. Thus, for example, Freundlich [117] recommended their use for reducing TiO_2 to metallic titanium. A silicocalcium alloy, whose composition corresponds approximately to $CaSi_2$, is used for the final deoxidation of some grades of steel [544]. According to GOST 4762-49 (All-Union State Standard) three grades of calcium-silicon are smelted (KaSi0, KaSi1, and KaSi2). The Ca + Si content of the best grade must be not less than 90%, of which the Ca should be not less than 31%, and there should be not more than 1.5% Al, 0.04% S, and 0.05% P; it should also contain up to 10% Fe, 1% C, and about 0.2% Mg. In the least pure grade (KaSi2), the Ca + Si content must be not less than 85% (of which the Ca should be not less than 23%). Calcium-silicon is smelted in shaft furnaces from quartzite, calcium carbide, and a mixture of coke and wood charcoal. The over-all electricity consumption (taking into account the preparation of CaC_2) is about 13,500 kw·hr [544]. As a result, calcium-silicon is an expensive material. The specific gravity of molten calcium-silicon is less than that of the slag and it is therefore necessary to use special pouring apparatus.

Moist air decomposes calcium-silicon. It is therefore stored and transported in sealed vessels.

Sr–Si SYSTEM. The Sr–Si system has not been studied completely. Jacobs (1900) reported that he had synthesized strontium disilicide, $SrSi_2$, from $SrCO_3$ and SiO_2 by reduction with carbon in an electric furnace and noted that this compound did not give "silicoacetylene" when reacted with HCl. However, he gave no fuller details. According to Wöhler and Schuff [118], the preparation of strontium silicides requires a higher temperature than that for calcium silicides. These authors prepared strontium disilicide in porcelain crucibles in an argon atmosphere at a temperature slightly above 1400° from a mixture of strontium and silicon. Strontium monosilicide, SrSi, was prepared by heating the same mixture for a short time at a temperature up to 1000°. At a higher temperature, the metal volatilized and it was difficult to synthesize the monosilicide (instead of the latter, the disilicide was obtained). Strontium silicides react with water very vigorously. When heated in air they give strontium silicates. Wöhler and Schuff also determined the heats of formation of the strontium silicides (see Table 2).

The information given on the Sr–Si system is very limited. Further systematic study of it is needed, using modern investigation methods.

The Sr–Si system and strontium silicides have no practical value as yet.

Ba–Si SYSTEM. Like the previous one, the Ba–Si system has not been investigated at all. Bradley [119] reported that he obtained barium disilicide, $BaSi_2$ by heating a mixture of $BaCO_3$, SiO_2, and carbon in an electric furnace. This compound is bluish white or steel gray and is decomposed by water with the evolution of hydrogen. When reacted with HCl, it gives silicon hydrides and silica.

Wöhler and Schuff [118] prepared a series of barium silicides analogously to the strontium silicides. The monosilicide, BaSi, was prepared by heating a mixture of barium and silicon to 1000° and the others by fusion at 400° in an argon atmosphere. These authors reported that they also isolated the following barium silicides: $BaSi_2$, $BaSi_3$, and even Ba_2Si_7 or $BaSi_4$. Barium silicide, containing free silicon, and barium polysilicides are stable to the action of moisture. The heats of formation of barium mono- and trisilicides reported by these authors are given in Table 2.

This is the extent of the information on the Ba–Si system at present and the system has no practical use as yet. A systematic investigation of this system using modern procedures is required.

Ra–Si SYSTEM. The Ra–Si system has not been studied at all. There are no data on radium silicides although in analogy with strontium and barium, they presumably exist.

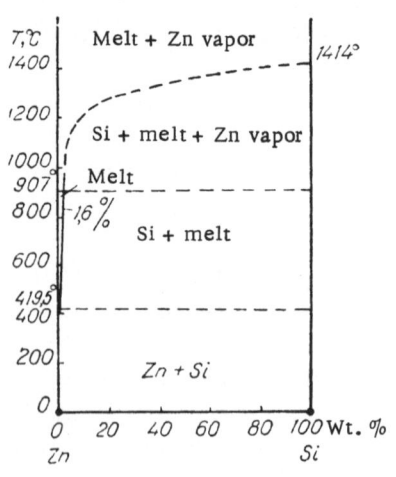

Fig. 22. Hypothetical phase diagram of the Zn–Si system based on the data of Moissan and Siemans (1904).

Systems Formed by Silicon with Elements of Group IIB

In contrast to the elements of the main subgroup of Group II, the elements of the secondary subgroup do not form silicides with silicon. There is some similarity between the behavior of these elements and that of the elements of the main subgroups of Groups III, IV, and V. This is evidently related to the structure of their electron shells and as a result, the bonds with silicon are weaker than those within the crystal lattice of the pure element.

The Zn–Si, Cd–Si, and Hg–Si systems have no practical value.

Zn–Si SYSTEM. St. Claire Deville and Caron (1857–1863) and Winkler (1864) established that all the silicon in a melt with zinc separated in the free state when the melt crystallized, without forming a silicide. Vigouroux [120] tried to synthesize zinc silicides in an electric furnace but obtained negative results. Moissan and Siemans [121] established that the solubility of silicon in zinc is as follows:

Temperature, °C	Solubility of silicon in zinc, %
600	0.06
650	0.15
730	0.57
850	1.62

Using this data and also the analogy with the structure of systems with silicon in other secondary subgroups, we have constructed a hypothetical phase diagram for the Zn–Si system, shown in Fig. 22.

Table 8

Crystal Lattice Parameters of the Zn–Si System

Material	Lattice parameters, A		
	Zn		Si
	a	c	a
Pure Si	—	—	5,4170
Pure Zn	2,6581	4.9341	—
Alloy of 50% Zn and 50% Si	2.6578	4.9353	5.4164

By investigating an alloy of zinc and silicon, annealed at 265°, Jette and Gebert found the crystal lattice parameters [89] given in Table 8.

Jette and Gebert concluded from their data that solid solutions are not formed in the Zn–Si system (more accurately: their concentration was less than could be detected by measuring the lattice parameters. – A.B.).

Cd–Si SYSTEM. The Cd–Si system has not been studied. Jette and Gebert [89] measured the crystal lattice parameters of a cadmium silicon alloy, which was heated slightly below the boiling point of cadmium (below 767°) and then annealed at 257°. The results of their determinations are given in Table 9.

Table 9

Crystal Lattice Parameters of the Cd–Si System

Material	Lattice parameters, A		
	Cd		Si
	a	c	a
Pure Si	—	—	5.4170
Pure Cd	2.9713	5.6046	—
Alloy of 50% Cd and 50% Si	2.9709	5.6058	5.4181

From the data obtained Jette and Gebert concluded that no chemical compounds or solid solutions with appreciable concentrations of the second component were formed in the Cd–Si system.

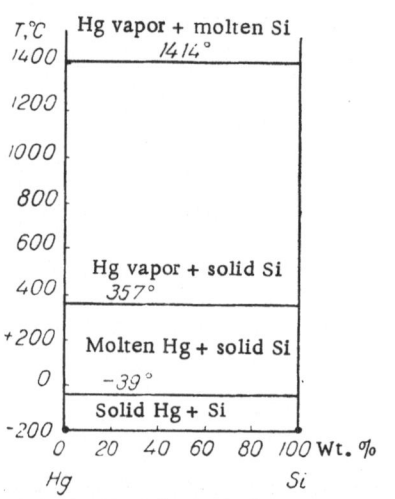

Fig. 23. Hypothetical phase diagram of the Hg–Si system.

Hg–Si SYSTEM. Winkler (1864) established that mercury does not react with silicon when the two are heated in sealed vessels. Guertler [122] reported that there was no reaction between these substances at normal temperatures. The phase diagram of the Hg–Si system should have the form shown in Fig. 23.

Systems Formed by Silicon with Elements of Group III

For various reasons it is impossible at present to make any general remarks on the systems formed by elements of Group III with silicon. Systems with rare earth elements (lanthanides) are examined separately.

B–Si SYSTEM. In 1900 Moissan and Stock [123] carried out experiments on the preparation of boron silicides (silicon borides) by fusing a mixture of boron and silicon with an electric current, as in the synthesis of carborundum. These investigators considered that they obtained two compounds: opaque black grains of B_6Si (spec. gravity 2.47) and the black, rhombic, slightly translucent, platelike crystals of B_3Si (spec. gravity 2.52); both compounds were electrically conducting. These data of Moissan and Stock have since been printed in handbooks and textbooks, though for almost 40 years no attempts were even made to repeat or improve the Moissan experiments. Only in 1938, Tone [124] tried to reprepare the compound of boron and silicon, but his experiments gave negative results.

In 1951 American investigators [125] heated two mixtures of boron and silicon (with 20 and 50 at. % of Si) for 20 minutes at 1725° but there were no indications that compounds of these elements were formed. The x-ray powder pictures of the preparations obtained had strong lines of elementary silicon and somewhat weaker lines that were assigned to boron.

Stavrolakis et al. [246] heated mixtures of boron and silicon at 1400° but did not obtain compounds. By means of hot pressing, they obtained dense, solid objects from these mixtures, which were stable to oxidation when heated and had a bending strength of 2810 kg/cm^2.

Boltaks [42] reported the use of solid solutions of boron in silicon with a B content of 1%, which had an electrical resistance a factor of 10,000 less than that of pure silicon.

Fuller and Ditzenberger [126] studied the diffusion of boron (from BCl_3) and phosphorus in silicon and found that the diffusion coefficient D depended on the absolute temperature T according to the following equation:

$$D = 0,001 \exp\left(-\frac{58\,000}{RT}\right).$$

The almost identical activation energies for the diffusion of the two elements in silicon (58,000 cal) and the value of the pre-exponential factor show that the diffusion mechanism was determined mainly by the diffusion of silicon itself. Fuller and Ditzenberger established by these experiments that the solubility of boron in silicon at 1250° is within the limits of $(3-6) \cdot 10^{20}$ at/cc, which corresponds to approximately 0.3 wt. % of B.

Samsonov and Latysheva [127, 128] recently reported that by hot pressing very pure mixtures of boron and silicon at 1600–1800° they obtained a single compound, B_3Si, regardless of the composition of the charge. These investigators reported that B_3Si crystallized in a tetragonal system (see Table 2) and not a rhombic one as Moissan had supposed. Its microhardness (5352 ±167 kg/mm^2 with a load of 30 g) is considerably greater than that of boron

(3340 kg/mm^2) or silicon (1808 kg/mm^2). The compound B$_6$Si was prepared by Samsonov and investigated by Zhuravlev [714]. According to Moissan, B$_6$Si and B$_3$Si have an exceptionally high chemical stability. Sulfuric acid decomposes them only on boiling. Nitric acid reacts with B$_6$Si, while molten alkalies react more vigorously with B$_3$Si. B$_6$Si and B$_3$Si are oxidized in air only when heated to a high temperature.

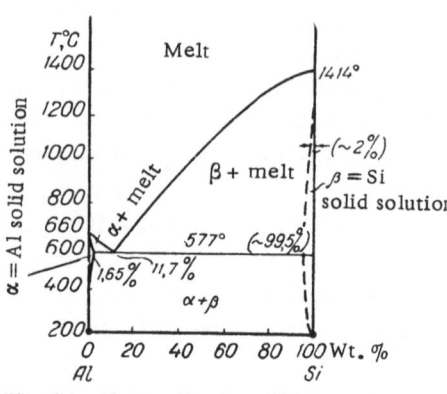

Fig. 24. Phase diagram of the Al—Si system according to Fraenkel (1908), Roberts (1914), Gwyer (1926), Phillips (1927), Broniewski and Smialowski (1932), and Spengler (1955).

Considering the properties of boron and silicon, one should assume the absence of low-melting eutectics in this system. Boron and silicon compounds, if such exist, should have high melting points (hypothetically of the order of 2000°). As the atomic radii of boron (0.91 A) and silicon (1.17 A) differ from each other by almost 29 and 22%, respectively, the formation of solid solutions by these elements with a large area of homogeneity can hardly occur.

Due to the great importance of refractory articles in industry, a detailed study of the B—Si system, and development of its phase diagram are required despite experimental difficulties.

Al—Si SYSTEM. As a result of the work of a series of investigators [129—134] (thermal analysis and study of microstructure), the phase diagram of the Al—Si system has been established and it is shown in Fig. 24. The solubility limit of aluminum in silicon is given according to [134].

As Fig. 24 shows, the Al—Si system contains no binary phases and, with the exception of narrow areas of solid solutions of silicon in aluminum and aluminum in silicon, it forms only a eutectic. The microhardness of Al—Si alloys was investigated by Savitskii and Baran [633].

The presence of a third component did not result in the formation of silicides containing aluminum as a rule, except in those cases where the latter partially replaced silicon. For example, such a compound is examined below in the section on chromium silicides.

Phragmen [637] investigated the Al–Cu–Mg–Mn–Fe–Si system in the region of alloys rich in aluminum and found in it the following silicon-containing phases: $Al_8Mg_3FeSi_6$–hexagonal, $a = 6.62$ kX, $c = 7.92$ kX, specific gravity 2.82; $AlCuMgSi$–hexagonal, $a = 10.30$ kX, $c = 4.04$ kX, specific gravity 2.79; $AlMnSi$–cubic, $a = 12.625$ kX, specific gravity 3.55; $AlFeSi$–cubic, $a = 12.523$ kX, specific gravity 3.61; $AlFeSi$–monoclinic, $a = b = 6.11$ kX, $c = 41.4$ kX, $\beta = 91°$, specific gravity 3.39; $AlFeSi$–tetragonal, $a = 6.11$ kX, $c = 9.46$ kX, specific gravity 3.43.

All these silicides were etched either weakly or not at all by acids, thus indicating quite considerable chemical stability.

It was stated above (see section on the Mg–Si system) that small amounts of magnesium and silicon form solid solutions in aluminum. Kuznetsov and Makarov [135] investigated such solid solutions by x-ray methods.

The Ca–Al–Si system is known to contain two ternary compounds [241], $Ca_3Al_6Si_2$ and $Ca_2Al_4Si_3$, though their properties have been studied little as yet.

The Al–Si system is very important in technology. Alloys of aluminum and silicon. Silumin (4.5 – 14% Si, 0.5% Mg, 0.5% Mn, up to 1% Fe, and sometimes copper and zinc), have high strength with a low specific gravity. They are used for casting. Aluminum alloys that are poorer in silicon (by several percents) are also widely used (Lautal, Aldrey, and some types of Duralumin).

An alloy of aluminum and silicon, aluminum-silicon, which is prepared by reducing kaolin and quartz with wood charcoal in electric furnaces [545] or from some coals [546, 547], is also used as a reducing agent and in the preparation of various alloys. According to technical conditions, aluminum-silicon must contain 25–40% Al, 50–60% Si, and not more than 0.05% S and P. In the best grades, the iron content should not exceed 5%.

Ga–Si SYSTEM. Klemm et al. [136] used thermoanalysis in vacuum to determine the phase diagram of the Ga–Si system (E. Hohmann). Samples containing a large amount of gallium were investigated in quartz vessels in vacuum. Alumina crucibles were used (once each) for silicon-rich alloys and the alloys were investigated in an argon atmosphere. In all, four alloys were investigated apart from pure gallium and silicon (5, 10, 20, and 40 at.% Si). The phase diagram obtained for the Ga–Si system is given in Fig. 25. This is a simple eutectic system without any peculiarities. The solubility of gallium in solid silicon was not investigated. The melting point of the eutectic is very close to that of pure gallium [139]. Its composition was not determined exactly, but calculations [139] show that the eutectic contains only very minute amounts of silicon.

Very pure silicon, containing only about 10 gallium atoms per million silicon atoms, was isolated from melts of the Ga–Si system. These alloys may be used for preparing very pure silicon monocrystals [710].

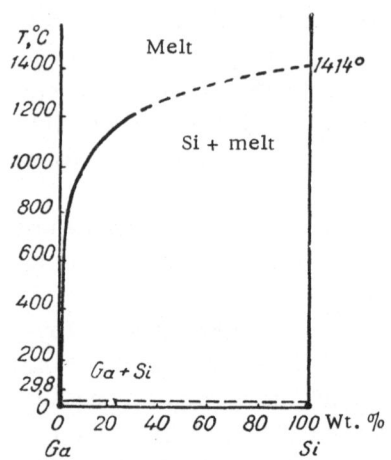

Fig. 25. Phase diagram of the Ga–Si system according to Hohmann (1948).

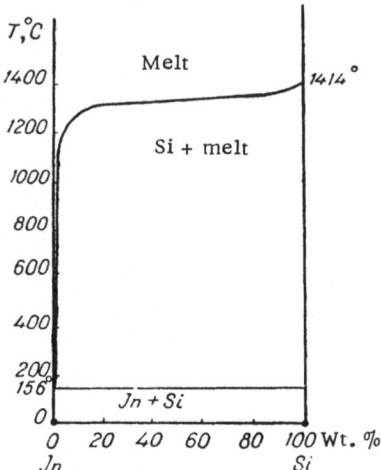

Fig. 26. Phase diagram of the In–Si system according to Klemm (1948).

In–Si SYSTEM. Klemm [136] also worked out the phase diagram of the In–Si system, which is shown in Fig. 26. 99.9% pure indium was used. Samples for plotting the thermograms were placed in corundum crucibles and heat treated in a hydrogen atmosphere. The crucibles cracked after one operation with alloys rich in silicon. In all, six alloys were investigated (with approximately 16, 20, 28, 58, 70, and 90 at.% Si), apart from the pure components.

Figure 26 shows that the In–Si system is a eutectic one. The shape of the liquidus curve indicates that in the region of silicon contents of 15–90 wt.%, there should apparently be a region of immiscible melts. However, no immiscibility of the melts was observed experimentally. The eutectic point in this system is very close to the melting point of indium, while the silicon content of the eutectic was very small according to calculations [139].

The melts of the In–Si system may be used for preparing very pure silicon monocrystals [710].

Tl–Si SYSTEM. The Tl–Si system was investigated indirectly by Tamaru [137], who used nine binary alloys and found that thallium did not fuse with silicon due to immiscibility in the liquid state. The phase diagram of the Tl–Si system is shown in Fig. 27.

As in the case of the two preceding systems, the Tl–Si system is not used for practical purposes.

An examination of the phase diagrams of the Al–Si, Ga–Si, In–Si, and Tl–Si systems shows that with an increase in atomic weight and, consequently, in the complexity of the electron shell, the solubility of silicon in the molten metal decreases. Such a phenomenon is also characteristic of the Zn–Si and Hg–Si systems described above.

Sc–Si AND Y–Si SYSTEMS. The Sc–Si and Y–Si systems have not been investigated as yet. In analogy with rare earth elements, which are examined after the description

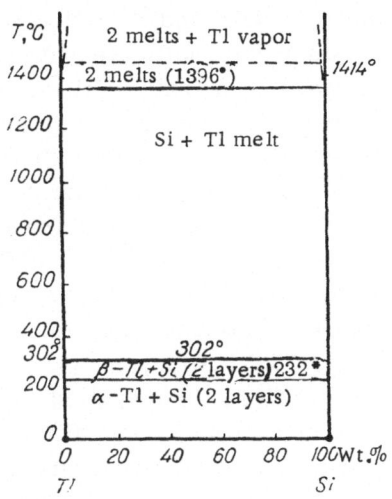

Fig. 27. Phase diagram of the Tl–Si system according to Tamaru (1909).

of systems of silicon with platinum group elements, there should presumably be silicides with the composition $MeSi_2$ ($ScSi_2$–55.6% Si and YSi_2–38.72% Si). Scandium silicide of unknown composition and an yttrium disilicide, YSi_2, with a specific gravity of 4.35 and a structure differing somewhat from that of α-ThSi$_2$, were recently obtained [613]. It is doubtful whether the Sc–Si and Y–Si systems will have any value other than a general theoretical interest as examples of systems with silicon. Data on systems formed by silicon with elements of the actinide group are described at the end of this section.

Systems Formed by Silicon with Elements of Group IVB

The same rules are observed in this subgroup in systems with silicon as for elements of the main subgroup of Group III. With an increase in the atomic weight and, consequently, in the atomic radius, the miscibility of the metals with silicon sharply decreases, even in the molten state, and practically complete immiscibility is reached with lead.

The practical value of systems of silicon with elements of the main subgroup of Group IV is so varied that it cannot be considered as a whole.

C–Si SYSTEM. Berzelius (1824) observed the formation of a binary compound of silicon and carbon during the reduction of potassium fluorosilicate with metallic potassium. Depre (1849) obtained a compound of silicon and carbon by reduction of silica with carbon by heating the latter with an electric current. However, Depre did not establish the composition of the compound obtained. Schützenberger [141] first obtained silicon carbide in a definite form by heating a mixture of one part of silicon with two parts of silica in a carbon crucible. The temperatures in Schutzenberger's experiments were comparatively low and therefore the silicon carbide he obtained appeared as an amorphous substance.

In 1891, Acheson [142], who was carrying out experiments on the artificial preparation of diamonds, heated a mixture of corundum contaminated by silica and carbon and obtained a new compound. He first took this compound to be aluminum carbide and gave it the name "carborundum" (from the words carbon and corundum). On establishing that this was a compound of carbon and silicon, he then obtained the same substance by heating a mixture of silica and carbon with sodium chloride added in an electric furnace. He patented this method and from then until the present time, it has been the main one for the preparation of carborundum.

Silicon carbide (carborundum) was obtained by Moissan [143] before Acheson, but the former published the results of his experiments later than Acheson. Moissan obtained silicon carbide SiC by different methods: 1) direct synthesis from the elements at a temperature of 1200–1400°; 2) crystallization from molten iron; 3) reduction of silica with carbon; 4) the action of silicon vapor on carbon; 5) reduction of silica with calcium carbide. Moissan [144] also described the silicon carbide found in nature (in the meteorite from Canyon Diabolo), which Kunz (1905) named "moissanite" in his honor.

Due to its exceptional hardness, carborundum began to be used for grinding soon after its discovery even though it cost up to 1800 roubles per kilogram [145]. However, the cost of carborundum fell after development of its industrial production in the USA in 1893, and then in Europe. Carborundum has been produced in the USSR since 1931 by methods developed by our own specialists.

The great development of the carborundum industry in connection with the high hardness, electroconductivity, and refractoriness of carborundum has evoked a huge amount of investigations and publications. The first collection of information on carborundum was given in the monograph by Fitz-Gerald [146].

Colson [147] described yet another silicon carbide with the composition SiC_2, which he obtained by heating silicon in ligroin vapor in a stream of hydrogen. However, further investigations established that the statements of Colson and other authors [154] were incorrect and it was shown that the C–Si system has only one chemical compound, namely carborundum, whose composition corresponds very accurately to the formula SiC. Pring [148] established that carborundum was formed from silicon and carbon at a temperature of even 1250° and according to the newest data [149], below 1150° also.

The formation of carborundum by reduction of silica with carbon even proceeds at temperatures slightly above 1200°, but at a low rate. Therefore, under industrial conditions, the reaction is carried out at a temperature of about 2000°. As the thermodynamic calculations of Kamentsev showed [150], this reaction proceeds according to the equation

$$SiO_2 + 3C = SiC + 2CO$$

without the intermediate formation of free silicon, as was supposed previously. It is possible that SiO is also formed as an intermediate product [155].

The heat of formation of carborundum from silicon and graphite is 26.7 kcal/mole at 25°. The entropy equals 3.935 entropy units and the heat capacity, C_p, according to Magnus [210], depends on the absolute temperature, T (273–1729°K), according to the equation

$$C_p = 8.89 + 2.91 \cdot 10^{-3}\, T - 2.84 \cdot 10^5\, T^{-2};\quad (2\%).$$

Analogous results were obtained by Immke and Kratzers [211].

For the reaction

$$Si + C \rightleftarrows SiC$$

the equilibrium constant $K = 4.90 \cdot 10^4$ at 1300° [158].

Thermodynamic data on black and green carborundum are also presented by Maksimenko and Polubelova [376].

Carborundum cannot be fused at atmospheric pressure. Lampen [151] established that carborundum begins to decompose at very high temperatures, but reported 2200° for this temperature. According to the latest data, hexagonal carborundum dissociates only at a temperature of about 2700° [152] and if it is not able to decompose, then a melt is formed also.

Despite the value of carborundum in technology, the phase diagram of the C–Si system has not been established up to the present time due to the very great experimental difficulties. Figure 28 shows the two probable variants of the phase diagram of the C–Si system

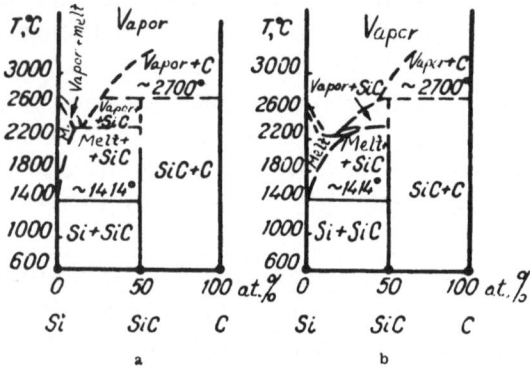

Fig. 28. Probable phase diagrams of the C–Si system according to Nowotny, Parthe, Kieffer, and Benesowsky (1954).

according to Nowotny et al. [153]. From analysis of residues after evaporation of various mixtures of silicon and carborundum, it was established that the following equilibrium occurs at atmospheric pressure:

vapor + carborundum ⇌ silicon-rich melt

Hence, it can be assumed that variant a of the phase diagram of the C–Si system (Fig. 28) is closer to reality than variant b.

X-ray and optical investigations showed that the crystal lattice parameters of silicon, graphite, and carborundum in preparations, fired at high temperatures, did not change. This established the very narrow range of homogeneity of the phases mentioned in the C–Si system.

It has been known for a long time that carborundum is characterized by the presence of several forms, of which one is cubic and the others hexagonal [156]. The formation and decomposition of carborundum was studied by Ruff [205] and its formation from silica and carbon, by Novikov also [207]. The effect of sesquioxides on the formation of carborundum was established by Kamentsev [206].

Taylor and Laider [149] thoroughly investigated the formation of carborundum from silicon and silica and established that only cubic carborundum was formed at temperatures up to 2000°. According to Baumann [157], cubic carborundum is formed from silicon and carbon at 525° in the presence of an aluminum–zinc melt.

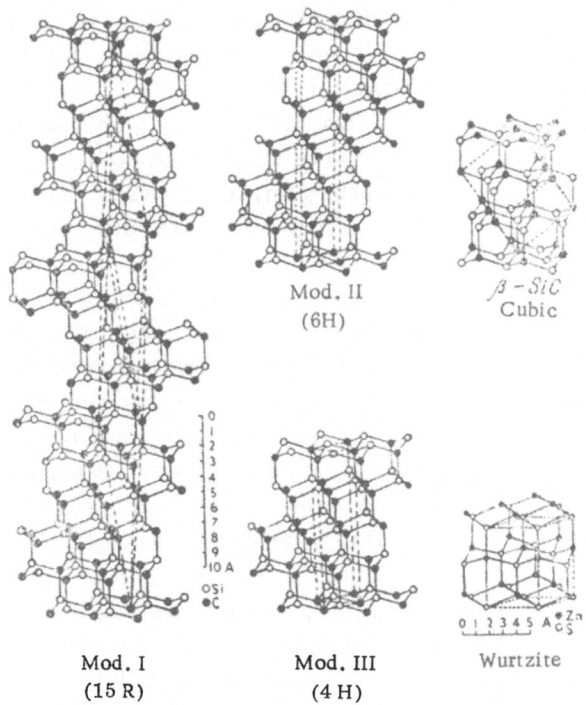

Fig. 29. Crystal lattices of some modifications of SiC and wurtzite according to Ott.

On the basis of thermodynamic data, Humphrey et al. [155] showed that the cubic modification of carborundum is stable up to 1800°K (1527°) and the hexagonal, at a higher temperature. According to Humphrey, the heat of combustion of cubic carborundum equals 7248.7 cal/mole and that of the hexagonal, 7276.1 cal/mole. The difference in the heat content of these carborundum modifications at 298.16°K equals 60 cal/mole, and in entropy, 0.03 entropy units. The change in volume in the change from β-SiC to the hexagonal α-form is very small (about 0.06%).

Baumann [157] found that cubic β-SiC changes monotropically into the hexagonal α-SiC. This conversion occurs in a few minutes at 2300°.

The cubic modification of barborundum has a sphalerite type structure (diamond). This structure of carborundum was investigated by Ott [156] and other authors [159] (Fig. 29).

The structures of a series of hexagonal forms of carborundum were studied by Ott, and the crystal lattice parameters determined more accurately by Bormann and Seyfarth [160].

After this there appeared a large number of publications on new forms of hexagonal carborundum, α-SiC (see for example [149, 161 – 175]), whose results are presented in Table 2.

The hexagonal forms of carborundum are derived from the cubic form by rotation of the tetrahedra at whose centers are layered atoms of silicon or carbon. All these forms of carborundum are characterized by the densest sphere packing of the atoms. All the silicon and carbon atoms are arranged on the three vertical axes of symmetry of the hexagonal cell. The number of similar axes of symmetry and the number of layers in a period along the hexagonal axis can serve as a classification index for the different hexagonal forms of carborundum.

Fig. 30. Structure of SiC-II, according to Belov (1947).

The main tetrahedra of silicon and carbon atoms in hexagonal carborundum are oriented parallel or antiparallel, forming blocks of similarly directed layers. This is readily seen from the example of the disposition of such tetrahedra in SiC-II, illustrated in Fig. 30 (according to N. V. Belov). The most stable carborundum structures have two or three layers or their combinations. It has recently been established that there are four-layer structures – carborundum VIII, 4.4, according to the symbols of Zhdanov and Minervina [170, 171] and in type 21R, 3·4 [171].

The elementary tetrahedra in carborundum are distorted. They are elongated along the c axis and compressed along the a axis. Taylor and Laider [149] found, for example, that for SiC-II the deviation from the ideal structure was -0.0007 A for the a parameter and $+0.0179$ A for c. All the structures of carborundum are polar.

The nomenclature of different hexagonal forms of carborundum was developed by Zhdanov [163] and Ramsdell [165], who based it on the layer packing of the tetrahedra. An examination of the numerous hexagonal forms of α-SiC led Zhdanov [166] to the conclusion that they were not all modifications of carborundum, but only kinetic formations, appearing due to regular dislocations during the growth of the crystals, as occurs, for example, in micas and feldspars or during the formation of defects in hexagonal cobalt [149]. According to Zhdanov, the basis of all the structures of hexagonal carborundum is SiC-II (symbol 3.3) and to a lesser extent, SiC-III (symbol 2.2). The other forms, with rare exception, are combinations of these forms. Taylor and Laider [149] consider the possibility of the existence of an even simpler structure for hexagonal carborundum of the wurtzite type (see Fig. 29), but it has not been observed for carborundum up to now.

Thus, strictly speaking, there are only two modifications of carborundum: cubic β-SiC and hexagonal α-SiC. The presence of many forms of α-SiC is explained by the ready formation of defects during the growth of carborundum crystals with a slight energy difference. The latter is caused by the identity of the dispositions of neighboring atoms. The differences appear only in the disposition of the atoms in the second and further coordination. Ramsdell and Kohn [171] proposed that the reason for the formation of numerous forms of α-SiC is the different size of the carborundum particles in the vapor state (different molecular weight. – A.B.), but they were unable to present any experimental data in favor of this. The more plausible conclusions of Soviet specialists (Ormont, Zhdanov, et al.) indicate that the reason for the formation of different forms of hexagonal carborundum is the presence of impurities. Thus, it was found [56] that at high temperatures and in the absence of boron, SiC-II was formed and in the presence of boron, SiC-III. Despite the hypotheses given on the effect of the vapor phase, Ramsdell and Kohn arrived at the same conclusion and stated that even the cubic form is frequently observed in industrial carborundum at crystal edges and peaks. This is apparently due to the effect of formation temperature.

At the present time, 18 hexagonal forms of carborundum are known. The largest cell contains 393 molecules of SiC.

According to the latest data [167], the communications of Japanese authors [175] on the discovery of α-SiC with $z = 594$ and $c \approx 1500$ A are incorrect (inaccuracies in the calculations of Laue diffraction patterns); these values must be reduced by a factor of two.

Ramsdell and Kohn [171] attempted to develop theoretical ideas on the coexistence of different forms of α-SiC. However, there are as yet no experimental data on conversions of hexagonal carborundum from one form to another.

Honjo [179] established the presence of two types of anomalous structures for carborundum and of these, one was caused by irregularity in the superposition of planes formed by silicon and carbon atoms.

In connection with the existence of such a large number of different forms of α-SiC for a very simple composition, very detailed investigations were made of the structures of the growth of its crystals [197]. For this purpose, in particular, a very thin layer of silver was deposited on the faces of carborundum monocrystals, making it possible to observe the fine structure of the faces under a microscope with side illumination. It was found [180] that the basal planes of α-SiC crystals have a left or right spiral structure. The first type has trigonal vicinal spirals along (0001) with indexing in class C_3. The second type has a hexagonal spiral with the index C_6. Vicinals were not found on the prismatic and pyramidal faces.

The growth spirals of α-SiC, in connection with the polytypical nature of the latter, were studied in particular detail by Verma [181]. According to Verma, three types of growth spirals are observed on α-SiC crystals: 1) elementary with a pitch approximately equal to the dimensions of the elementary cell, 2) those derived from dislocations of increasing size, whose pitch equals the short dimension of the elementary cell, 3) interlaced spirals, whose step height is a fraction of a dimension of the elementary cell.

The step height of some spirals reaches 35 A.

The elementary spirals on the basal faces of carborundum arise from Frenkel defects of the lattice and have a definite crystallographic direction. Polygonal spirals are formed around an entrained inclusion and are found at some distance from it. Verma established the presence of interaction of the growth front of carborundum crystals and growth spirals and also their dependence on the number of lattice defects. The greatest density of measured defects on the given faces of hexagonal carborundum was characterized by about 100,000 spiral whorls per square centimeter of surface. Interferometric investigations showed that the step height of elementary growth spirals on the hexagonal faces of carborundum corresponded to the dimensions of the elementary cell in the rhombohedral form, as can be seen from Table 10.

Table 10

Characteristics of the Step Height of Growth Spirals of α-SiC

Type of α-SiC	Index according to Ramsdell	Size of cell edge, A		Step height of growth spirals, A
		in hexagonal axes	in rhombo-hedral axes	
II	6H	15,1	—	15 ± 2
I	15R	37.95	12,73	12 ± 2
VI	33R	82.94	27,7	28 ± 2

On some faces of α-SiC crystals, oriented "overgrowths" (for example, in the form of right pyramids) were observed beside growth spirals, which were obtained during the production of carborundum. Hexagonal spots, voids, and other forms of defects in the normal structure were observed in the centers of some spirals. The well-illustrated investigation of Verma confirms previous hypotheses on the reasons for the polytypical nature of α-SiC being dislocations in the crystal growth.

Buckley [182] found that the width and depth of some growth spirals and formations similar to them on the [0001] faces of α-SiC were of an order of several thousand times the elementary cell dimensions. According to Buckley, such formations could also occur due to twisting of the elementary layers during the growth of carborundum crystals. Vand [242] attempted to use the theory of dislocations of the carborundum lattice to explain its polytypical nature.

Structural dislocations undoubtedly cause not only the polytypical nature of α-SiC, but also some variation in its technical properties. Therefore, they deserve further study. The hypotheses

presented above on the reasons for the polytypical nature of carborundum were examined by Frank [183], who demonstrated the
formation of dislocations in the carborundum lattice during the
growth of crystals.

Of 150 α-SiC crystals investigated by Jagodzinski [184], 62 had
one-dimensional defects. Jagodzinski found that the cubic modification of carborundum, β-SiC, was the only stable one and that
the other α-SiC forms were the result of dislocation formation.
For large-sized crystals, the hexagonal forms were found to be
more stable than the thermodynamically stable cubic structure
of carborundum.

Table 11

Phase Composition of Technical Carborundum [176]

Carborundum	Content of various types of SiC, %		
	β-SiC	SiC-II	SiC-III
Soviet "Extra".............	—	100	—
Soviet black	—	60	40
From experimental melt No. 23 .	—	—	100
American commercial	25	60	15
American cubic............	100	—	—

Ramsdell found that coarse crystals of carborundum were predominantly SiC-II and also SiC-I and SiC-III. The SiC-IV and SiC-V
forms were present as unique crystals. Technical carborundum
consisted mainly of SiC-II and also 87 R, 51 R, and 33 R (VII, V,
and VI).

Zhdanov, Minervina, and Nevzorova [176] developed a method
of phase analysis of carborundum, giving an accuracy of 10–20%,
and showed that the so-called amorphous carborundum was almost
pure β-SiC (cubic). The results of phase analysis of various samples of technical carborundum are presented in Table 11 (according
to Zhdanov).

Carborundum SiC-I was not detected in the preparations Zhdanov
investigated, though it is frequently encountered in the form of separate, coarse crystals. The reason for this was apparently the very
small over-all amount of it in the preparations investigated.

Up to now there have been no investigations on the dependence
of the technical properties of carborundum on its phase composition, although such a dependence must exist.

Hexagonal carborundum (apparently SiC-II) has the following optical properties (sodium light): $Ne = 2.697$, $No = 2.654$, $Ne - No = 0.043$, optically positive and strong dispersion [185]. The presence of very thin films of SiO_2 causes iridescence effects with some carborundum crystals.

Carborundum is practically involatile when heated to 1730° [24]. On the basis of the data presented by M. V. Kamentsev [150], the vapor pressure of Si + SiC over carborundum at higher temperatures approximately obeys the relation

$$\lg P = -\frac{50420}{T} + 20.294,$$

where P is the vapor pressure in mm Hg and T is the absolute temperature.

Hence, the activation energy for carborundum evaporation equals 58.4 kcal/mole. The dissociation of carborundum in the gas phase is slight. At a temperature of 2300–2700°, the free silicon content of the gaseous products over carborundum is not more than 5–10% [150].

According to Khantverger [145], the coefficient of linear thermal expansion of polycrystalline hexagonal carborundum at a temperature of 100° is $6.58 \cdot 10^{-6}$ and at a temperature of 900°, only $2.98 \cdot 10^{-6}$. For refractory articles of recrystallized carborundum (refrax), the average coefficient of linear thermal expansion is about $5.0 \cdot 10^{-6}$ for 25–1400°. The thermal conductivity of polycrystalline carborundum equals 0.015–0.023 cal/sec · cm · deg [145]. The anisotropy of the thermal conductivity of carborundum crystals has not been studied as yet.

The change in the electrical resistance of polycrystalline hexagonal carborundum with temperature is illustrated by Fig. 31. Carborundum is characterized by a negative temperature coefficient of electrical resistance. The electroconductivity of carborundum has been the subject of many investigations (see for example [186–193, 236, 237]). From measurements of the electroconductivity of monocrystals between 77 and 1400°K, Busch and Labhart [187] found that yellow and green carborundums (type n) were semiconductors with excess electrons, whereas black carborundum (type p; 0.2% Al and Ca and 0.1% Mg) had an electron deficit, which confirmed the previously established [237] difference in the nature of the electroconductivity of green and black carborundums.

Ivanov and Pruzhinina-Granovskaya [377] showed that the presence of Fe^{3+} and Cr^{3+} ions increased the electroconductivity of carborundum. The addition of aluminum gave good results in the preparation of electrotechnical carborundum.

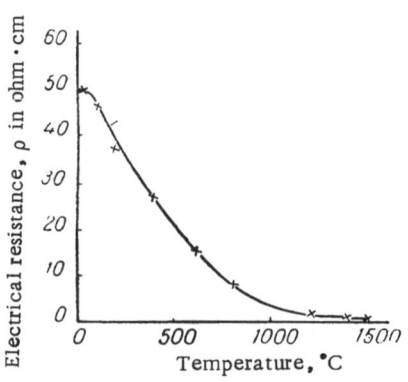

Fig. 31. Electrical resistance of polycrystalline samples of carborundum in relation to temperature.

By new investigations on very pure carborundum, obtained by decomposition of tetrachlorosilane and toluene vapor on graphite platelets, Kendall [193] determined the conductivity activation energy, which changed quite considerably with temperature.

Carborundum synthesized in the presence of metal chloride vapor may acquire different types of conductivity, n or p. The results obtained show that the conductivity of carborundum cannot be explained by either deviations from the stoichiometric composition (excess or deficit of silicon atoms) or by the presence of foreign atoms. However, it is known that very pure carborundum will not conduct an electric current. Dudnik and Pruzhinina-Granovskaya [190] showed that the duration of intermittent processes in black carborundum does not correspond to ideas on electron conductivity, but should rather be ascribed to the diffusion of ions from the semiconductor into the "exhaustion layer." At the present time, the mechanism of carborundum electroconductivity has not been explained satisfactorily [191, 192] and neither has its negative temperature coefficient of electrical resistance. Further investigations of this problem are necessary.

The oxidizability of carborundum in air at high temperatures is of very great importance and limits its use as a refractory and an electrical heating material. The oxidation of this substance in air begins practically only at temperatures above 1000°. Brodsky and Gubicciotti [194] established that at 1015–1020° the oxidation of polycrystalline carborundum proceeds according to the equation

$$Q = K \lg(1 + a\tau),$$

where Q is the amount of oxygen taken, K and a are constants and τ is time. Figure 32 shows the results of measuring the oxidizability of carborundum at 1400°, based on data of Kamentsev [150]. Processing of these data leads to an equation similar to that of Brodsky and Gubicciotti,

$$Q = 12 \lg(1 + 0.04\tau) \approx 0.7\sqrt{\tau},$$

i.e., the oxidation of carborundum is approximately proportional to the square root of the time. The temperature dependence in this case is probably described by an exponent.

The formation of an SiO_2 film on carborundum crystals has a shielding effect, but due to the presence of microfissures in it, it cannot stop oxidation.

At high temperatures, water vapor decomposes carborundum according to the reaction

$$SiC + 2H_2O = SiO_2 + CH_4.$$

Sulfur dioxide reacts with it more vigorously than water [195] and water vapor with air, even more vigorously.

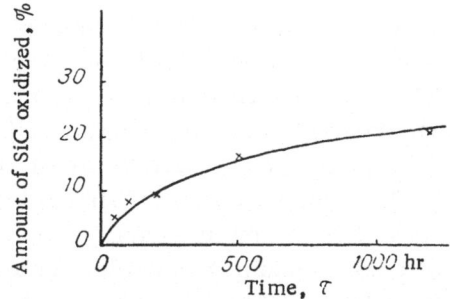

Fig. 32. Oxidation of polycrystalline carborundum at 1400° according to Kamentsev (1950). $y \approx 0.7\sqrt{\tau}$, or $y = 12 \lg(1 + 0.04\tau)$, where y is the percent of carborundum oxidized and τ is time.

Nitrogen does not react with carborundum at 1100°, while chlorine reacts with it even at low temperatures. Here the course of the reaction changes with temperature [15]:

at 100°

$$SiC + 2Cl_2 = SiCl_4 + C,$$

at 1000°

$$SiC + 4Cl_2 = SiCl_4 + CCl_4.$$

Sulfuric and hydrofluoric acids do not react with carborundum, whereas orthophosphoric acid decomposes it.

Molten alkali hydroxides and carbonates decompose carborundum with the formation of silicates.

$$SiC + 2NaOH + 2O_2 = Na_2SiO_3 + H_2O + CO_2,$$
$$SiC + Na_2CO_3 + 2O_2 = Na_2SiO_3 + 2CO_2,$$
$$SiC + 2Na_2O_2 + O_2 = Na_2SiO_3 + Na_2CO_3.$$

The lead compounds PbO and $PbCrO_4$ decompose carborundum with heating. Carborundum reacts with ferric oxide at temperatures above 1000°. Under these conditions, Fe_2O_3 accelerates the oxidation of carborundum in air due to its oxygen, but slows the oxidation at 1200° due to the formation of protective films of an iron-containing silicate melt.

Carborundum reduces many oxides to the metals at high temperatures. Magnesium and calcium oxides, sodium silicates, alkali sulfates, borax, and cryolite strongly decompose carborundum on heating [196].

In connection with the use of carborundum as an abrasive, mixtures of it with boron carbide have been studied. It was established [124] that carborundum and B_4C mix readily in the molten state and form a simple eutectic system with an SiC content of approximately 35%, though the melting point of the eutectic was not determined.

The separation of SiC in skeletal forms in the B_4C mass made it possible to obtain sharp grains. When the melt of boron carbide and carborundum was cooled, cubic α-SiC and not hexagonal β-SiC crystallized. Boron carbide may be used as a binder for carborundum articles. Good results are obtained in the preparation of the latter by hot pressing. Tone [124] obtained nonporous articles of this type, containing up to 80% SiC. Carborundum is used in combination with boron nitride and other substances for the manufacture of firing electrodes and ignitron rectifiers.

The B–C–Si system is of exceptional interest, not only for the preparation of abrasives, but also in refractory technology. Combinations of carborundum with borides have been proposed [666] for rocket parts. A detailed study of the B–C–Si system with the development of its phase diagrams is therefore necessary. The use of a silicon nitride binder has been recommended [225, 665] for the manufacture of special carborundum refractories.

Carborundum is widely used in technology for the manufacture of electrical heaters (Globar and Silit resistors, etc.), protective disks in electrotechnics (thyrite), as abrasives, and also as a highly heat-conducting refractory. The use of carborundum as a refractory material was examined particularly by Everhard [198]. For use as a heat-resistant material, carborundum is bound with silicon. Such material, manufactured by the electrothermal method of the firm "Carborundum and K°" (USA), under the name of "Durhy," possesses the following properties [27]: a tensile strength at room temperature equal to 4570 kg/cm^2, at 800°, 4220 kg/cm^2, and at 1320°, 2320 kg/cm^2; its coefficient of linear thermal expansion equals $4 \cdot 10^{-6}$; its heat conductivity is 0.04 cal/cm · sec · deg at 850°. Where transition metals such as chromium are used as binders for carborundum, the latter decomposes.

Aluminum carbide, Al_4C_3, is decomposed by silicon [544]. The formation of ternary compounds in the Al–C–Si system is improbable. Hence it follows that Al_4C_3 and aluminum are in equilibrium with carborundum. This may be employed for practical purposes in the application of carborundum refractories in the metallurgy of aluminum and its alloys and also for the metallization of the contacts (leads) of carborundum electrical heaters.

The occurrence of the reaction

$$CaC_2 + Si = CaSi + 2C.$$

was established [544] in the Ca–C–Si system.

Hence we must conclude that metallic calcium is not in equilibrium with carborundum and will decompose it on heating with the formation of calcium silicides and carbon or CaC_2.

Carborundum is decomposed in the presence of iron.

$$SiC + Fe = FeSi + C.$$

The free energy of this reaction at high temperatures is determined by the equation [544]

$$\Delta F_T^0 = 9900 - 9.14 \, T.$$

From the equation it follows that carborundum is unstable in the presence of iron above 810°. The use of carborundum refractories for prolonged service in contact with iron above this temperature is thermodynamically impossible, but short-term use is possible due to the kinetic aspect of the process.

High quality crucibles for melting metals are manufactured from carborundum with a carbon binder [199]. Combinations of carborundum and silicides of transition metals are of interest [41, 641]. Molybdenum-containing silicate binders are recommended for decreasing the oxidation of carborundum at high temperatures [196]. Data on the Mo–C–Si system are presented below in the section on molybdenum silicides.

The hot pressing of carborundum in graphite molds at very high temperatures and pressures (2530° and 700 kg/cm² for α-SiC and 2370° and 422 kg/cm² for β-SiC) makes it possible to prepare articles with a porosity of 1–5% [648]. The preliminary synthesis of β-SiC at 1450° is recommended. Excess silicon and carbon are removed by flotation and acid treatment of the powdered material.

The addition of 1% Al is used to raise the stability at high temperatures. β-SiC changes into the α-form during hot pressing at temperatures above 2000°.

Articles of hot-pressed carborundum with a porosity of 1% have a fracture strength of 4900 kg/cm^2 at 1371°. Hot-pressed articles of carborundum and molybdenum (40 vol. %) have a bending strength of 5062 kg/cm^2 at 1000° and very good resistance to air up to 1370°.

It has been proposed that dense articles of hot-pressed carborundum could find application as the operating parts of polishing-wheel cutters, in the manufacture of crucibles, as linings for orifices in wire drawing (dies), and as the uncooled parts of rocket motors.

Mixtures of steatite (60–93%) and carborundum (7–40%) have been recommended for the manufacture of turbine blades [667].

The exceptionally valuable properties of carborundum and peculiarities in the structure of its monocrystals indicate a need for

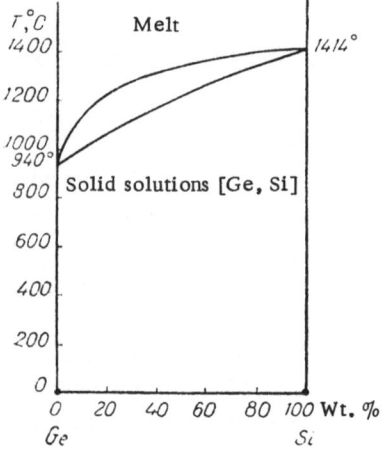

Fig. 33. Phase diagram of the Ge–Si system according to Stöhr and Klemm (1939).

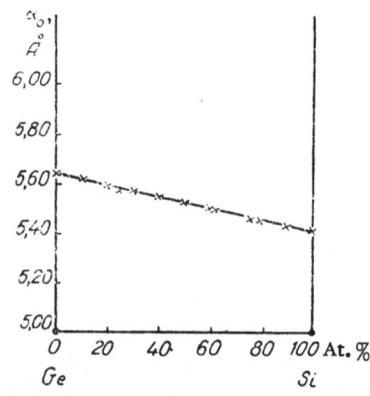

Fig. 34. Size of the elementary cell edge (a_0) of germanium–silicon solid solutions in relation to composition.

further thorough study of the Si–C system and polycomponent systems including it (especially the Si–C–B system) and also of carborundum itself. The technical importance of this material will undoubtedly increase.

Ge–Si SYSTEM. The identical structures of the crystal lattices (diamond type) of germanium and silicon and the very small dif-

ference in the size of the elementary cell (about 4%) are the reasons
for the formation of a continuous series of solid solutions (replacement) of these elements. The phase diagram of the Ge–Si system
(Fig. 33), worked out experimentally by Stöhr and Klemm [200]
from an investigation of eight melts, shows that the formation of
the simplest type of solid solution occurs here.

The liquidus curve of the Ge–Si system was obtained by means
of thermal analysis. The solidus curve necessitated holding samples at high temperatures for very long periods (several weeks,
and sometimes up to six months).

X-ray investigations confirmed the formation of solid solutions
of germanium and silicon.

The crystal lattice constants of silicon–germanium solid solutions changed almost linearly with composition (Fig. 34).

A recent check of the phase diagram of the Ge–Si system [201]
confirmed the results of Stöhr and Klemm.

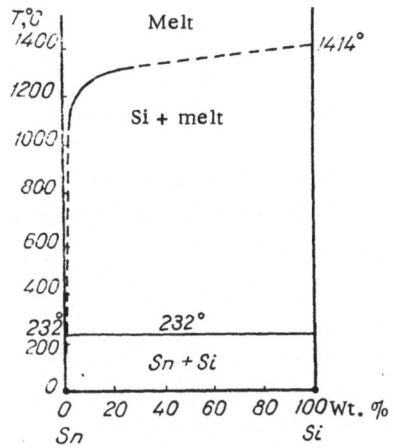

Fig. 35. Phase diagram of the Sn–Si
system according to Tamaru (1909).

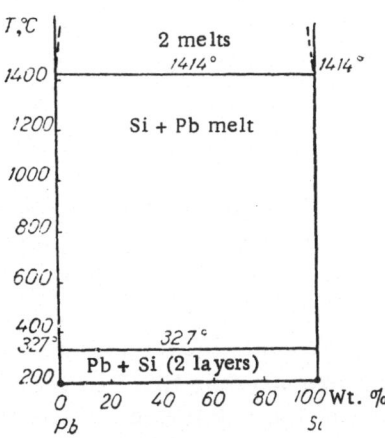

Fig. 36. Phase diagram of the Pb–Si
system according to Moissan and
Siemens (1904) and Tamaru (1909).

A thermodynamic investigation of the solubility of silicon in
germanium showed [139] that the solutions formed in this system
were close to ideal. The presence of anomalous heat effects on the
cooling curves of germanium–silicon melts (at 930° according to
Stöhr and Klemm and 938.7–940.4° according to Hassion, Goss, and
Trumbore) are caused by diffusion phenomena in the solid phase.

The electrical resistance of germanium–silicon solid solutions (at 300–800°K) was studied by Levitas [202].

The Ge–Si system is of importance in semiconductor technology.

Sn–Si SYSTEM. Vigouroux [203] established that alloys of tin and silicon consist of pure components and this was confirmed by Tamaru [204] by thermal analysis and microscopic investigations. Tamaru investigated 11 binary alloys and established the phase diagram of the Sn–Si system (Fig. 35), from which it follows that the liquidus curve rises very steeply toward silicon. Tin and silicon mix in the molten state in any concentration.

Table 12

Crystal Lattice Parameters in Tin–Silicon Alloys [89]

Material	Annealing temperature, °C	Lattice parameters, A		
		Sn		Si
		a	c	a
Pure Si	—	—	—	5.4170
Pure Sn	—	5.8167	3.1728	—
Alloy with 2% Si	595	5.8156	3.1744	5.4165
» » 50% Si . . .	150	5.8178	3.1728	5.4184
» » 80% Si	595	5.8151	3.1748	5.4174

Thurmond [139] found that the melting point of the eutectic in the Sn–Si system at a very low silicon content must be lower than the melting point of tin by 0.001° according to calculation. Therefore, Tamaru was not able to establish either the melting point or the composition of the eutectic in the Sn–Si system.

Jette and Gebert [89] established that in alloys of tin and silicon, prepared at 1520° in vacuum and annealed, the lattice parameters of the components were almost unchanged (Table 12).

The table shows that tin and silicon do not form appreciable solid solutions.

The applied value of the Sn–Si system is slight as yet.

Pb–Si SYSTEM. Even in the Nineteenth Century (St. Claire Deville, 1857; Winkler, 1864, and others), it was established that lead and silicon do not mix. Vigouroux [203] and later Moissan [121] showed that silicon dissolves in lead to only a very slight extent: 0.02% at 1250°, 0.15% at 1400°, 0.21% at 1450°, and 0.78% at 1550°. They also established that the Pb–Si system does not form chemical compounds.

Tamaru [204], working with technical silicon (about 92% Si), investigated seven binary alloys by means of thermal analysis and established the phase diagram of the Pb–Si system, which is presented in Fig. 36 and which is characterized by the presence of a very large region of immiscible melts (at almost all concentrations).

Jette and Gebert [89] showed that in an alloy of 50% Pb and 50% Si, annealed at 130–150°, the lattice parameters of lead and silicon (4.9375–4.9382 and 5.4156–5.4160 A, respectively) hardly differed from those of the pure substances (lead – 4.9385 A, silicon – 5.4170 A), indicating the very low concentration of silicon in the solid solution in the Pb–Si system. Silicon has the valence 2.5 in solid solution in lead, copper, and other metals with a type A structure [244].

The practical value of the Pb–Si system is very low.

Systems formed by silicon with transition elements of Groups IV, V, and VI are described separately as they have much in common.

Systems Formed by Silicon with Elements of Group VB

The rules given for binary systems of elements from the main subgroup of Group IV with silicon also apply to the systems examined in this section. However, the silicon-nitrogen bond is considerably weaker than the silicon–carbon bond and this appears, for example, in its rupture during the action of water on nitrogen-containing organosilicon compounds and in the dissociation of silicon nitride, Si_3N_4, at a considerably lower temperature (1900°) than carborundum. The silicon phosphorus bond is even weaker. Due to the considerable difference in the atomic radii and the greater ease of electron addition, arsenic does not form solid solutions with silicon like germanium, but gives chemical compounds. In their behavior toward silicon, antimony and bismuth are completely analogous to tin and lead, respectively.

N–Si SYSTEM. The first experiments on the preparation of silicon nitrides were carried out starting from the middle of the last century (Bailman, 1844; St. Claire Deville and Wöhler, 1859; Geuther, 1865; Schützenberger, 1879; Titherley, 1894). The products obtained by the direct reaction of silicon and nitrogen at high temperatures or from SiO_2 and alkali amides were ascribed the following different compositions without adequate grounds: SiN, Si_2N_3, Si_3N_4, SiN_2, and others (sometimes Si_2N_5, SiN_3, and Si_3N are still being mentioned).

Blix and Wirbelauer [212] established that the reaction of $SiCl_4$ with NH_3 formed a white mass with the composition $Si(NH_2)_4$, which, on heating to 1200–1300° in a stream of nitrogen, was converted into silicon nitride with the evolution of ammonia by the scheme

$$Si(NH_2)_4 \rightarrow Si(NH)_2 \rightarrow S_2N_3H \rightarrow Si_3N_4.$$

Vigouroux [213] established that nitrogen does not react with silicon even at 1000°. Weiss and Engelhardt [214] studied the action of nitrogen on silicon at 1120, 1220, 1320, and 1420°. They found that when the materials were held at the maximum firing temperature for half an hour, the reaction between silicon and nitrogen proceeded at a measurable rate at 1240–1300°. Under these conditions there was only 0.01% of N in the reaction product, while the maximum content was up to 10% when the firing time was increased.

The silicon nitrides obtained by these authors were impure (traces of free silicon and SiO_2). Depending on the time at 1300–1400°, nitrides were obtained to which were ascribed the compositions Si_2N_3, SiN, and Si_3N_4. The nitrides obtained were white or light gray in color and had a fibrous structure. Weiss and Engelhardt considered Si_3N_4 to be "normal" silicon nitride, into which the others were converted by prolonged heating in a stream of nitrogen. The specific gravities of the silicon nitrides obtained by Weiss and Engelhardt are presented in Table 2. According to the latest data [233], the specific gravity of Si_3N_4 determined pyknometrically is not correct. We also consider that the specific gravity of Si_2N_3, determined as 3.64 by these authors, is doubtful. If such a nitride exists, then its specific weight apparently should be considerably less.

By repeating the experiments of St. Claire Deville (heating silicon in a carbon crucible in a stream of nitrogen), Weiss and Engelhardt established that silicon nitride was not formed thus, but a compound with the composition Si_3C_3N.

Funk [215] studied the formation of silicon nitride by the action of nitrogen on very finely divided silicon, isolated from its alloy with aluminum. The binding of nitrogen could be detected even at 1100–1200°. A nitride with the composition Si_3N_4 was formed after 10 minutes at 1450°. Funk ascribed its gray color to contamination with silicon carbonitride.

Schwarz and Sexauer [216] reported that when silicon triimide was heated to 490° in dry nitrogen, polymerized silicocyanogen was formed,

$$Si_2N_3H_3 = (Si_2N_2)_x + NH_3,$$

but they not only did not provide any proof that the solid reaction product obtained had the composition indicated, but even reported the possibility of forming a hydrogen-containing compound of the silicam type.

Frank and Louis [116], who studied the nitridation of $CaSi_2$, considered that a silicon nitride was thus formed, but they did not study its composition. However, the importance of the following equilibrium was noted:

$$Si_2N_2 \rightleftarrows SiN_2 + Si.$$

Finley [49] reported the silicide Si_3N (85.7% Si), but also without any proof. It was noted that the compound Si_2N_3 was formed by the action of nitrogen on carbazosilicon at high temperatures.

$$Si_2C_2N + N_2 = Si_2N_3 + 2C.$$

Carbazosilicon may be obtained by heating SiC in a stream of nitrogen.

Friedrich and Sittig [217] heated a mixture of silica and carbon in a stream of nitrogen with a few percent of hydrogen at 1500° and obtained silicon nitride with the composition Si_3N_4.

$$3SiO_2 + 6C + 2N_2 = Si_3N_4 + 6CO.$$

Where 10% Fe_2O_3 was added to the charge, the reaction temperature was reduced to 1250–1300°. The iron could then be removed by treatment with hydrochloric acid, in which Si_3N_4 is insoluble, to leave the nitride as a white powder.

Information on the preparation of other compounds of silicon with nitrogen is even more scanty. The phase diagram of the N–Si system has not been studied at all, though it is not only of theoretical, but also of practical value. The silicon nitride with the composition Si_3N_4 has been studied in the greatest detail.

The thermodynamic characteristics of silicon nitride, Si_3N_4, were studied by Matignon [218], Roth and Börger [219], Satch [220], Kelly [221], Weibke and Kubaschewsky [222], and others. The heat of formation of this compound is given as the following:

Author	$-\Delta H_{298°}^\circ$, kcal/mole
Matignon (1913)	230.0
Roth and Börger (1937)	168.0
Satch (1937)	163.0
Weibke and Kubaschewsky (1943).	160.0 ± 15
Quill (1956)	179.2

Apparently, the values determined by Satch and also Weibke and Kubaschewsky are the most reliable ones for the heat of formation of Si_3N_4. According to Kelly [221], the entropy of Si_3N_4 equals 23.0 ± 2.5 entropy units and according to Quill [24], 20.4 entropy units. The most reliable entropy value is apparently that of Kelly.

The dependence of the heat capacity of this compound, C_p, on the absolute temperature, T, based on Satch's measurements [220], is given by the following equation for temperatures of 273–860°K:

$$C_p = 15.70 + 29.4 \cdot 10^{-3}T - 6.3 \cdot 10^{-6}T^2, \quad (5\%).$$

The heat of the reaction

$$Si_3N_{4\,cryst.} + 5O_{2\,gas} = 3SiO_{2\,cryst.} + 4NO_{gas,}$$

is 349.4 kcal/mole according to Nasini and Cavallini (1934).

Silicon nitride is completely dissociated into the component elements in the gas phase [24, 226].

Hincke and Brantley [223] investigated high-temperature equilibria between Si_3N_4, silicon, and nitrogen, and established that the partial pressure of nitrogen, P, at temperatures of 1600–1800°K (the partial vapor pressure of silicon is low at these temperatures in comparison with nitrogen) in relation to the absolute temperature, T, may be determined from the equation

$$\lg P = 8.575 - 19\,323 \cdot T^{-1} \text{ [atm.]}.$$

The free energy, ΔF, of the reaction

$$Si_3N_4 = 3Si + 2N_2$$

can be found from the equation

$$\Delta F = 176\,300 - 78.35\,T.$$

Extrapolation gives $\Delta F = 0$ at 2250°K (1977°), which corresponds to a dissociation pressure equal to one atmosphere. As is known [217], Si_3N_4 evaporates at 1900° under a pressure of one atmosphere.

Friedrich and Sittig [217] established that Si_3N_4 crystallizes in a noncubic system. Leslie, Carrol, and Fisher [224] investigated the structure of the silicon nitride Si_3N_4 isolated from specially nitrided steel with a high silicon content (up to 3.2%). Chemical, x-ray, and electron diffraction methods were used for the investigation. Chemical analysis of the silicon nitride samples showed that their composition corresponded to the formula $Si_3N_{3.94}$, i.e., very close to theoretical. The characteristics of the powder diagram of Si_3N_4 according to these authors are presented in Table 13.

Silicon nitride, Si_3N_4, and germanium nitride, Ge_3N_4, were found to belong to the rhombic system and not the rhombohedral, as was proposed previously for Ge_3N_4 [227]. The parameters found for the Si_3N_4 crystal lattice are presented in Table 2. According to more recent data. there are two modifications of Si_3N_4: rhombic and rhombohedral.

Table 13

Characteristics of Si_3N_4 X-Ray Powder Diagram [224]

d observed	I	d observed	I	d observed	I
6.88	Very weak	2.305	Strong	1.507	Weak
4.28	Strong	2.267	Weak	1.485	Average-strong
3.856	Average	2.232	»	1.436	Average
3.344	»	2.149	Average	1.416	»
3.107	Very weak	2.074	Strong	1.405	Average-weak
2.873	Strong	1.933	Weak	1.353	Average
2.801	Very weak	1.881	»	1.320	Weak
2.651	» »	1.859	»	1.309	»
2.583	Strong	1.800	»	1.301	Average-weak
2.529	»	1.764	Average	1.295	Weak
2.483	»	1.747	Very weak	1.260	Less weak
2.427	»	1.594	Average	1.225	Weak

The two nitrides, Si_3N_4 and Ge_3N_4, are isomorphous. A structure of the Ge_3N_4 type with a rhombohedral treatment is presented in Fig. 37. It is actually slightly deformed and, as was shown, is rhombic. This type of structure, close to phenacite, is explained by loose packing of the atoms (there are "channels" in the crystal lattice), which is the reason for the comparatively low specific gravity found for Si_3N_4 (see Table 2) in the latest investigations [225].

Since Si_3N_4 and Ge_3N_4 monocrystals have not been investigated up to now, the development of their structures must be considered approximate.

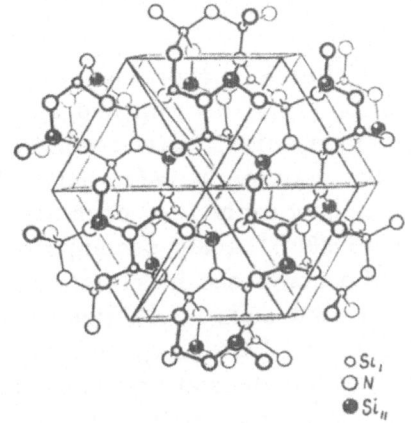

○ Si_I
○ N
● Si_{II}

Fig. 37. Crystal lattice of Si_3N_4 with a rhombohedral treatment.

In connection with the peculiarities in the structure of the crystal lattice, the thermal expansion coefficient of Si_3N_4 is low: $2.46 \cdot 10^{-6}$ (20–1000°). The thermal conductivity of a polycrystalline sample of Si_3N_4 (its porosity is not known. – A.B.) over the temperature range 200–1300° was found to be 1.34 technical units or 0.00373 cal/cm · sec · deg [225]. Friedrich and Sittig [217] considered that Si_3N_4 does not conduct an electric current. According to more recent data [225], the electrical resistance of polycrystalline Si_3N_4 at room temperature is 1426 ohm · cm, which is considerably less than corundum, for example.

According to the determinations of N. V. Gul'ko (All-Union Scientific Research Institute on Refractories – VNIIO, 1956), the following optical constants were found for a sample of silicon nitride synthesized by O. M. Margulis (VNIIO): $Ng = 2.18, Np = 2.10$. The purity of the preparation obtained was not investigated.

The shear modulus of polycrystalline samples of Si_3N_4 is 112,500–140,600 kg/mm^2 at room temperature and 112,500 kg/mm^2 at 1000°; the compression strength at room temperature equals 5060–6330 kg/cm^2 [225]. The strength characteristics of Si_3N_4 change very little with an increase in temperature up to 1000°. This silicon nitride is not ductile and has a considerable hardness.

The great strength and insignificant thermal expansion make it possible to prepare comparatively heat-resistant parts from it.

In old papers it was reported that Si_3N_4 is decomposed by water.

$$Si_3N_4 + 6H_2O = 3SiO_2 + 4NH_3.$$

However, according to the latest data, this silicon nitride is not only not decomposed by water, but not even by most of the common acids, apart from HF. Old data on the instability toward water apparently refer not to Si_3N_4, but to intermediate products of considerably less chemical stability. Water vapor (investigated up to 800°) hardly decomposes Si_3N_4 [214]. Concentrated sulfuric acid decomposes it only partially.

Fused caustic alkalies decompose Si_3N_4 with the evolution of ammonia. This compound reacts very vigorously with lead oxides and chromate and this is used for its chemical analysis. This silicon nitride is stable to gases containing H_2S. Molten aluminum, lead, tin, and zinc do not decompose Si_3N_4. Magnesium reacts with it to only a small extent, but copper reacts more vigorously. Silicon nitride is comparatively stable to molten steel. The oxidation resistance of Si_3N_4 [225] and other refractory materials at 1200° is as follows:

Material	Increase in weight after 2 hours, mg/cm^2
Carborundum bound with silicon nitride...............	2.5
Silicon nitride	5.0
Titanium diboride	10.0
Hot-pressed titanium carbide..	42.5

These data show that silicon nitride has a comparatively satisfactory stability toward air oxidation at high temperatures, though it is considerably less than that of carborundum.

These properties of Si_3N_4 make it possible to use it as a refractory material, especially for the manufacture of protective covers for immersion thermocouples used in metallurgy. According to Collins and Gerby [225], such covers will stand 50 cycles in the metallurgy of nonferrous alloys and up to 45 cycles in a Martin furnace, while alumina thermocouple covers stand a maximum of 24 and an average of 10-12 cycles. According to the same data, in aluminum production, Si_3N_4 thermocouple covers were unchanged after 3000 hours operation at a temperature of about 930°. Therefore, despite preparation difficulties, such parts are of practical interest.

Silicon nitride reacts with transition metals of Groups IV, V, and VI according to the schemes [634]

$$Si_3N_4 + Me \rightarrow \text{metal nitride} + Si,$$

$$Si_3N_4 + Me \rightarrow Me_5Si_3 + N_2.$$

Only molybdenum does not react with it at comparatively low temperatures, but this requires checking by special experiments.

Articles of Si_3N_4 with the addition of carborundum (less than $1/3$) are recommended as abrasives [668].

This compound was found to be present in silicon steels [671, 676].

The N-Si system deserves systematic study with the use of modern synthesis methods and investigation of the reaction products obtained. The phase diagram of the N-Si system should be determined also.

P–Si SYSTEM.   The P–Si system has hardly been investigated at all. Berzelius (1843) considered that phosphorus and silicon do not give chemical compounds. However, Biltz [228] and his co-workers (Hartmann, Wrigge, and Wiechmann) showed that a silicon phosphide with the composition SiP was formed in this system and was analogous to GeP to some extent.

In Biltz' experiments silicon did not react with phosphorus in a sealed tube at 700° under the pressure of phosphorus vapor. A further increase in temperature was dangerous due to the sharp increase in the vapor pressure of phosphorus. Therefore, nonuniform heating of a sealed quartz vessel was used. Silicon was in the hot part (at 1000°) and phosphorus in the cool part (400°). The phosphorus vapor pressure was reasonable in this case. Under such conditions a yellowish brown, friable silicon phosphide without a metallic luster was obtained in 89% yield. With an increase in temperature in the hot part of the tube to 1150–1180° and in the cooler part to 450–480°, the silicon phosphide yield was 95%. Oxygen-containing compounds were absent from the product. Chemical analysis of this product showed that it corresponded to the formula SiP. This compound evolved all the bound phosphorus in vacuum at 1000–1100°. According to tensimetric investigations, this process began at 1026° and was measurable quantitatively at 1055°. The phosphorus vapor pressure over silicon phosphide in relation to temperature is shown

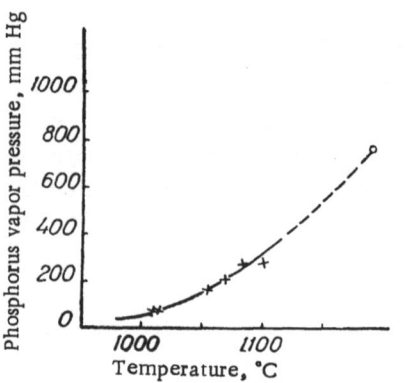

Fig. 38. Phosphorus vapor pressure over SiP in relation to temperature according to Biltz (1938).

by Fig. 38. It reached a pressure of 1 atm at 1140°, while for GeP, this is reached at 580°. This dependence is expressed approximately by the equation

$$\lg P = 13.78 - \frac{15\,400}{T} \text{ [mm Hg]},$$

where P is the phosphorus vapor pressure over SiP and T is the absolute temperature.

The melting point of silicon phosphide is above 1100°, but due to its volatility, its melting point is not reached at normal pressure.

Judging by the evolution of phosphorus, SiP is not the lowest silicon phosphide. The last 0.2 g-at. of phosphorus is evolved only at low pressure. Biltz considered that this could also be the result of solution of phosphorus in silicon. However, measurement of the solubility of phosphorus in silicon by Fuller and Ditzenberger [126] showed that it was only 1.3 wt.% at 1250° and decreased sharply with a fall in temperature. Therefore Biltz' observations are more likely to indicate the formation of another silicon phosphide, possibly with a composition close to Si_5P_2.

X-ray investigation of silicon phosphide, SiP, confirmed its individuality. The heat of formation of this compound equals 15.0 kcal/mole.

Silicon phosphide is hydrolyzed noticeably by boiling water with the formation of phosphine. A study of the hydrolysis process shows that the formula Si_3P_4 is also possible for silicon phosphide. However, in investigating it, Biltz and his co-workers preferred the simpler formula SiP, derived from chemical analysis. Evidently further elucidation by means of structural investigation methods is required.

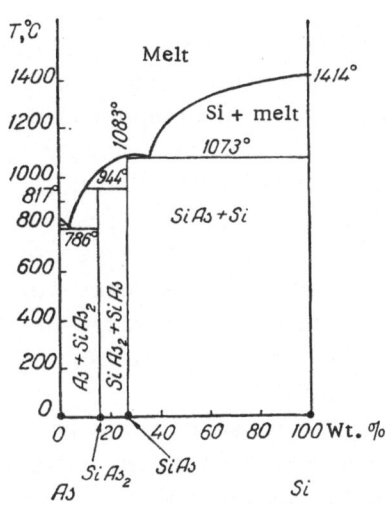

Fig. 39. Phase diagram of the As—Si system according to Klemm and Pirscher (1941).

Silicon phosphide does not react with dry oxygen, even on heating, and its surface only changes slightly. When it was heated in a stream of dry air, the weight loss was up to 4%. Apparently films of oxygen and phosphorus-containing compounds of silicon are formed on the surface of silicon phosphide crystals and protect it from further oxidation. Silicon phosphide reacts vigorously with chlorine.

The diffusion of phosphorus in silicon [126], as with boron, gives the equation

$$D = 0.001 \exp\left(-\frac{58\,000}{RT}\right),$$

where D is the diffusion coefficient; R is the gas constant; T is the absolute temperature (1270–1570°K).

The accuracy of this equation was checked at 970° also by means of radioactive phosphorus.

Solid solutions of phosphorus in silicon are used for the manufacture of detectors and semiconductor amplifiers [42]. With the addition of 0.001% P to silicon, the electrical resistance of the latter falls by a factor of 100 and with 1% P, by 10,000.

Considering the practical importance of the P–Si system, its detailed study and development of the phase diagram are desirable.

As–Si SYSTEM. Winkler (1864) carried out unsuccessful experiments on the preparation of silicon arsenides. Although Winkler considered the existence of the compound Si_6As possible, he could not obtain it either by heating arsenic with silicon or by treating the latter with arsenic vapor or H_3As. It was therefore thought [28] that chemical compounds did not exist in the As–Si system. However, Klemm, and Pirscher [229] studied the phase diagram of the As–Si system, refuted this opinion and obtained two silicon arsenides, $SiAs_2$ and SiAs, whose properties are presented in Table 2. X-ray powder pictures confirmed the existence of these arsenides. Their crystal structures were not established. It was only reported that there were a large number of lines on the x-ray powder pictures, indicating the complexity of the lattice structures of $SiAs_2$ and SiAs.

The phase diagram of the As–Si system is presented in Fig. 39 and in view of the simplicity of its structure, no special comments are required.

It was established by x-ray investigations that silicon arsenides dissolve in arsenic and silicon in only very small amounts, which remain unmeasured.

Both silicon arsenides are very stable chemically. There were no noticeable changes in these arsenides when they were boiled in water, hydrochloric acid, and KOH solution, but they decomposed when heated with NaClO and concentrated HNO_3. Part of the arsenic was evolved from $SiAs_2$ when the latter was heated in a porcelain crucible over a gas burner, but SiAs remained unchanged (the weight loss was less than 1% in half an hour).

The As–Si system and silicon arsenides have no practical use as yet.

Sb–Si SYSTEM. The first investigation of the Sb–Si system by Vigourous (1896) are now only of historical interest. The phase diagram of this system was worked out by Williams [230], who investigated in detail 14 binary alloys of very pure antimony and 98.1% silicon and also a series of supplementary ones. Williams established that there are no chemical compounds in this system. Therefore, reports on the presence of antimony silicides in contemporary literature [16] lack foundation. Antimony and silicon mix completely

Table 14

Crystal Lattice Parameters of Alloys of the Sb–Si System

Material	Annealing temperature, °C	Lattice parameters, A		
		Sb (in hexagonal coordinates)		Si
		a	c	a
Pure Si	—	—	—	5,4170
Pure Sb	—	4,295	11,247	—
Alloy with 40% Si . .	595	4,296	11,247	(Poor lines for measurement)
» » 50% Si . .	580	4,297	11,243	5,4176
» » 80% Si . .	595	4,296	11,250	5,4182

in the molten state. The absence of inclusions of silicon in antimony, visible under a microscope, when its content is less than 0.3 wt.% and of antimony in silicon, when the content of the former is 0–1 wt.% led Williams to conclude that solid solutions are formed at these concentrations. By x-ray determinations of the crystal lattice parameters, Jette and Gebert [89] showed that solid solutions were practically absent from the Sb–Si system. In any case, the solubility of silicon in antimony is less than 0.5% as can be seen from the constancy of its crystal lattice parameters presented in Table 14.

The phase diagram of the Sb–Si system according to Williams [230] with corrections according to the experimental results of Jette and Gebert [89] is presented in Fig. 40.

Thurmond [139] showed that the melting point of the eutectic in the Sb–Si system should be approximately 0.4° below the melting point of antimony and have an Si content of about 0.1%.

The Sb–Si system has not been used for practical purposes as yet.

Bi–Si SYSTEM. Vigouroux (1896) established that there are no chemical compounds in the Bi–Si system and this was confirmed by Williams [230]. Information on the existence of bismuth silicides, which is sometimes encountered in reviews, lacks foundation.

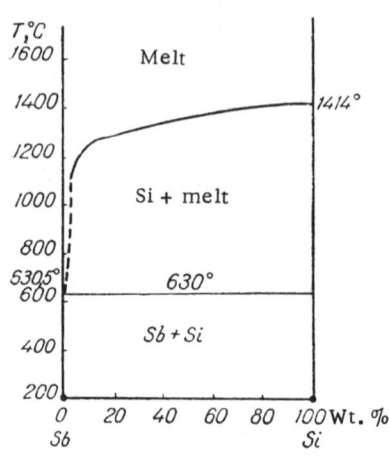
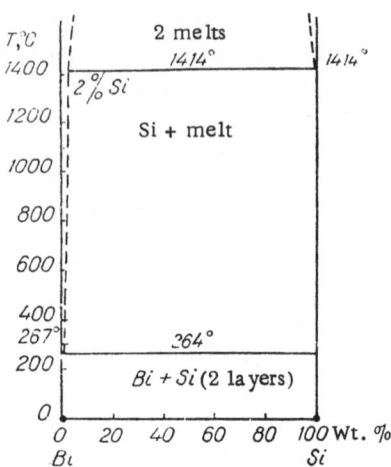

Fig. 40. Phase diagram of the Sb–Si system according to Williams (1907).

Fig. 41. Phase diagram of the Bi–Si system according to Williams (1907).

Williams [230] made a detailed study of nine binary alloys of the Bi–Si system and a series of supplementary ones. It was thus established that two melts were formed in the concentration range of 2–100% Si. Williams' supposition that a solid solution of silicon in bismuth was formed, which was based on the absence of inclusions of silicon in bismuth, visible under a microscope, at an Si content of up to 0.8%, was disproved by Jette and Gebert [89]. These investigators measured the crystal lattice parameters of bismuth and silicon in an alloy, containing 50% Si, and annealed at 240°. They found: for bismuth $a = 4.534$ A, $c = 11.830$ A (for pure bismuth $a = 4.535$ A, $c = 11.836$ A), for silicon $a = 5.4161$ A (for pure silicon $a = 5.4170$ A). Considering these data, the phase diagram of the Bi–Si system is presented in Fig. 41.

Williams found that the addition of silicon lowered the melting point of bismuth by approximately 3°. He did not establish the composition of the eutectic, but it was evident that the silicon concentration in it was very low.

The Bi–Si system has not acquired practical value as yet.

The systems formed by silicon with transition metals of Group V are described below.

Systems Formed by Silicon with Elements of Group VIB

Despite the accessibility of the starting components, systems formed by silicon with elements of the main subgroup of Group VI have hardly been studied at all up to the present. Even the Si–O system, which is the basis of a series of important industrial processes (especially in metallurgy), has been studied very little. Only recently, due to the discovery of a fibrous form of silica, has it been possible to find the presence of general characteristics in the structure of the compounds SiX_2, where X is O, S, Se, or Te. The low stability of compounds of the type SiX is also general.

The sharply expressed electronegative character of these elements is retained in their compounds with silicon. The melting points of the compounds SiX_2 fall sharply with the change from oxygen to sulfur, selenium, and tellurium. Compounds of silicon with sulfur, selenium, and tellurium are formed with a very great increase in volume (see Table 2), which indicates their low stability.

Si–O SYSTEM. The phase diagram of the Si–O system has not been established as yet. Therefore, only separate phases can be considered.

Investigations in a high-temperature x-ray camera showed that oxygen does not form appreciable solid solutions in silicon.

The silicon oxide poorest in oxygen is SiO. Winkler proposed the existence of such oxides of silicon in 1890. However, this was demonstrated experimentally only at the beginning of the Twentieth Century by Patter [247], who patented the process of manufacturing silicon monoxide, "Monox," for the preparation of pigments, abrasives, and lubricating materials. Flusin [248] established that "Monox" is a thermal and electrical insulator. Bonhoeffer [249], Saper [279], Jevancs [280], and others identified silicon monoxide in the gas phase spectrographically and then it was studied by many investigators [250, 251].

Silicon monoxide is obtained by evaporation of a mixture of silica and silicon in vacuum (not more than 10^{-4} mm Hg) at a temperature above 1250–1300° [41, 247, 268–270, 272] or in a helium atmosphere [41].

$$SiO_2 + Si = 2SiO.$$

The formation of silicon monoxide by this reaction begins at even 1100°. This reaction is used for the laboratory preparation of silicon monoxide by condensation of its vapor on cold parts of the apparatus and also for the preparation of films on mirrors. For this purpose cathode sputtering in vacuum (10^{-5} mm Hg) is also used and this yields a film of silicon and silicon monoxide [272].

When chamotte mixed with silicon is heated to 1450° in vacuum, silicon monoxide vapor is also obtained [254]

$$Al_2O_3 \cdot 2SiO_2 + 2Si = Al_2O_3 + 4SiO.$$

Silicon monoxide is contained in sublimates, formed by heating silicates in a reducing medium [266]. The volatility of a mixture of an Al–Si alloy and silica is explained by the formation of silicon monoxide [254].

Hydrogen and carbon reduce silica to silicon monoxide at temperatures over 1100°.

$$SiO_2 + H_2 = SiO + H_2O,$$
$$SiO_2 + C = SiO + CO.$$

Silicon monoxide is formed by the reaction of silicon with a series of metal oxides. This compound may also be obtained at 1300° in a CO_2 atmosphere at a pressure of about 0.01 mm Hg by the reaction

$$SiC + 2CO_2 = SiO + 3CO.$$

Gel'd and his co-workers [257] showed that reducing agents for silica could be arranged in the following decreasing order with respect to their action: gaseous silicon, solid carbon, liquid silicon, solid silicon, molten ferrosilicon, solid ferrosilicon, and carborundum. The rates of reduction of silica in vacuum at 1130–1370° by silicon and ferrosilicon are almost identical, while those for reduction by carborundum and carbon are considerably less. Silicon monoxide is contained in metallurgical slags formed under strongly reducing conditions and also in sublimates from electric steel melting furnaces [264]. Silicon monoxide was not found in the condensates of blast furnace vapor from furnaces melting silicon [267], but instead, silicon and SiC were present.

Gel'd and Esin [264] consider that silicon monoxide has Si–Si and Si–O–Si bonds of the type

$$\begin{array}{c} | \quad\quad | \quad\quad | \\ -Si-O-Si-O-Si- \\ | \quad\quad | \quad\quad | \\ -Si-O-Si-O-Si- \\ | \quad\quad | \quad\quad | \\ -Si-O-Si-O-Si- \\ | \quad\quad | \quad\quad | \end{array}$$

Investigation of the stability of SiO showed [265, 271, 272, 274, 277], that this compound decomposed according to the following equation at temperatures below approximately 1200°:

$$2SiO = Si + SiO_2.$$

Depending on the conditions, amorphous SiO_2, α-cristobalite, α-quartz, or mixtures of them were thus formed. Brewer and Edwards [277] established that the lower limit for the existence of silicon monoxide is a temperature of 1180° (1450°K). It melts at a temperature above 1700°.

Due to the difficulties of obtaining pure preparations, the heats of formation of silicon monoxide presented are very different [269, 270, 277]. The most accurate value is apparently $\Delta H_{298°}^{\circ} = -21,411 \pm \pm 570$ cal/mole for solid silicon and gaseous oxygen. The dissociation energy of this compound to the gaseous atomic elements is $169,300 \pm 3100$ cal/mole. Its heat capacity at 338°K equals 9.19 ± 0.16 cal/deg·mole. Gel'd and his co-workers [259, 263, 265] showed that silicon monoxide may exist at 500° in melts containing about 85% SiO.

The vapor pressure P (in atmospheres) of silicon monoxide in relation to the absolute temperature, T, may be found from the following equation according to Gel'd and Kochnev [256]:

$$\lg P_{SiO} = -\frac{16\,750}{T} + 1.75\,T + 1.9.$$

A silicon monoxide vapor pressure of 1 mm Hg is reached at 1325° and 1 atm at 1880°. Its heat of sublimation is 76,619 cal/mole. SiO is a monomer in the vapor state. Silicon monoxide and oxygen are obtained when silica is evaporated [277]. In the gas phase, SiO and SiO_2 are in equilibrium at 2000° [275].

Inuzuka [252] investigated the structure of the silicon monoxide crystal lattice and established its space group (see Table 2). In the SiO lattice, oxygen atoms surround silicon atoms in the form of irregular tetrahedra.

X-ray investigations of Biltz [253], Zintl [254], and Baumann [255] showed that in their solid preparations, silicon monoxide decomposed to silicon and SiO_2. Beletskii and Rappoport [267] considered that Inuzuka did not actually investigate silicon monoxide, but a mixture of $Si+SiC$. Only by heating a mixture of SiO_2 and carbon or silicon in vacuum at temperatures above 1800° were Beletskii and Rappoport able to obtain silicon monoxide with $a = 5.16$ A. Hoch and Jonston [273] determined the lattice parameters of this oxide at 25 and 1300° and hence the linear thermal expansion coefficient of this compound was found to be $4.6 \cdot 10^{-6}$. The lattice parameter of silicon monoxide was also determined by Gel'd and Popel' [265] and also by Jacobs [274]. Its refractive index, according to Gel'd and Popel', $N = 1.975-2.020$ and according to Beletskii and Rappoport, $N = 1.92-1.94$, so that the average value is 1.98. Hence, the specific refraction energy K is as follows:

$$K = \frac{N-1}{\gamma} = 0.439,$$

where γ is the specific gravity (Gladstone and Dale's rule). Consequently, the specific refraction energy of silicon monoxide ($K = 0.439$) is considerably greater than that of SiO_2 ($K = 0.207$). As stated above, silicon monoxide does not conduct an electric current.

In contrast to silicon, silicon monoxide is completely soluble in HF. Solid silicon monoxide reacts slowly with atmospheric oxygen at room temperature. Complete oxidation is not achieved at 500°. It is pyrophoric in a finely divided state and burns to SiO_2 in air, producing a flame. Water vapor reacts noticeably with SiO even at 500°.

$$SiO + H_2O = SiO_2 + H_2.$$

Silicon monoxide is slowly oxidized in a CO_2 atmosphere at 400°. Chlorine reacts with it at 800°.

$$2SiO + 2Cl_2 = SiCl_4 + SiO_2.$$

Sulfur dioxide also reacts with it at this temperature. SiO reacts with calcined dolomite and lime according to the equation [254]

$$MgO \cdot CaO + CaO + SiO = Ca_2SiO_4 + Mg.$$

A reaction corresponding to the following equation occurs in hot alkali.

$$SiO + 2NaOH = Na_2SiO_3 + H_2.$$

Silicon monoxide is used for the preparation of pigments and for mirror coating.

Despite numerous investigations, the existence of silicon monoxide was again questioned recently [278], especially with regard to the crystalline form of SiO. It is probable, however, that the preparation of the latter is very difficult, especially as a result of the low stability of silicon monoxide in the solid phase. The existence of analogs of this compound (for example, SiS) supplies additional confirmation of the existence of silicon monoxide, which must be investigated further.

A study of the products from the thermal decomposition of silicon monoxide in helium led Weber and Hessinger [41] to the conclusion that this was a new compound with the composition Si_3O_4 (43.3 wt.% O) or Si_2O_3 (46.1 wt.% O). Hönigschmidt had pointed out the probability of the existence of SI_2O_3 in 1909, but without suitable proof. Wagner and Pines [281] heated silicon oxyhydride (99.5% purity) at about 900°. Electron microscope investigations of the compound obtained, which corresponded to Si_2O_3 in composition, showed that it was present in the form of small spherical particles. Their melting point was found to be equal to 1635°. X-ray investigation showed that the substance obtained contained fine lines of silica, but was mainly amorphous. Silicon sesquioxide dissolved completely in aqueous hydrofluoric acid and did not change in weight when heated to 500° in a chlorine atmosphere, indicating the absence of free silicon from it. After fusion in an alumina crucible in an argon atmosphere at 1650°, a mixture of silicon and silica was formed from Si_2O_3. The use of Si_2O_3 for vacuum deposition of mirrors, like SiO, has been proposed.

Evidently data on Si_2O_3 have a very preliminary character and as yet, there is no conclusive proof of its existence. It is possible that Si_2O_3 is a mixture of SiO and SiO_2.

Consequently, the compound Si_2O_3 is not presented on the hypothetical phase diagram for the Si–O system put forward by Wagner and Pines and the data they obtained is used only for defining the position of the liquidus curve more accurately (Fig. 42).

Shishakov [282] considered that clays and pozzuolanes contain two-dimensional crystals with the composition Si_2O_5. However, they cannot be considered phases of the Si–O system as is the case with the "two-dimensional" cristobalite crystals of Nieuwenkamp [293]. The highest oxide of silicon is silica, SiO_2, which is the only binary compound of silicon, apart from carborundum, encountered in nature. The literature on silica is exceptionally voluminous [6, 7, 8, 10, 12,

15, 16, 17, 18, 20], and it is not necessary to attempt to cover all that is known on this problem but it is sufficient to examine only the main physicochemical aspects and new data.

As is known, the investigation of Fenner [292], which was carried out in 1913, is the classical work on the silica system. As a result

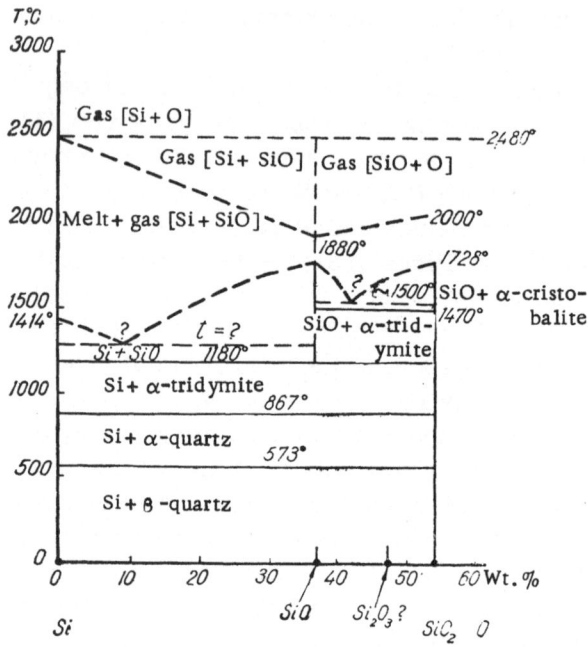

Fig. 42. Hypothetical phase diagram of the Si—O system.

of this investigation and minor individual refinements in further work by other authors, it was established that there are three types of silica modifications: quartz (two modifications), tridymite (three modifications) and cristobalite (two modifications), with high rates of phase conversion within the limits of each type and very low rates for the conversion of one type to another and then only in the presence of a melt or of water vapor (or other volatile substances). The phase conversion temperatures were also established and these were found to be a whole range in some cases (especially for $\beta \rightleftarrows \alpha$-cristobalite), which produced various hypotheses on the reasons for this phenomenon.

The thermodynamic characteristics of the main low-temperature forms of silica studied are presented in Table 15, where new data on the heats of formation reported by Humphry and King [322] are presented for quartz and cristobalite and the generally accepted older data are given for the other forms and in brackets for the two indicated.

Table 15

Thermodynamic Characteristics of Low-Temperature Forms of Silica

Form of SiO$_2$	$\Delta H^{\circ}_{298^{\circ}}$, kcal/mole	$S^{\circ}_{298^{\circ}}$, cal/deg·mole	$\Delta F^{\circ}_{298^{\circ}}$, kcal/mole
β-Quartz.	−210.26 (−205.4)	10.00	−192.4
β-Cristobalite	−209.33 (−203.0)	10.19	−192.1
α-Tridymite	−204.8	10.36	−191.9
"Quartz" glass	−202.5	11.20	- 190.9

Fenner proposed that chalcedony was a particular modification of silica. However, it is now known that chalcedony consists of particles of quartz or cristobalite, which are sometimes oriented [325]. Therefore, in the latest silica conversion schemes chalcedony does not stand out particularly.

Fenner reported that quartz or tridymite were not obtained under any conditions in the absence of a liquid phase. Cristobalite alone was always formed in this case. The appearance of silica modifications which were unstable (metastable) for the given temperature was explained for a long time on the basis of Ostwald's rule, which essentially contributed nothing to the understanding of the nature of the phenomenon.

The application of x-ray structural analysis made it possible to determine the structure of the various silica modifications and shed light on the nature of their conversions. It was established that the basis of the structure of all silica modifications and silicates are [SiO$_4$] tetrahedra and, with the exception of the recently discovered fibrous modification of silica, these are linked through the vertices. The structure of β-quartz (low-temperature) was investigated by Bragg and Gibbs [294], P'ei-Hsiu-Wei [295], Machatschki [296], and others. This structure is illustrated in Fig. 43 and the lattice dimensions and type are given in Table 2. The bond in quartz and tridymite is mainly of a homeopolar character.

The structure of β-quartz is characterized by the presence of spiral chains of [SiO₄] tetrahedra along the hexagonal axis and here the tetrahedra are somewhat deformed. With heating in a certain temperature range, the pitch of the chain spirals contracts and this is probably the reason for the negative expansion of α-quartz along the hexagonal axis [300]. The expansion coefficients of quartz increase sharply (Fig. 44) close to the conversion temperature (573°).

Jay [326] showed that β-quartz (low-temperature) shows considerable anisotropy with respect

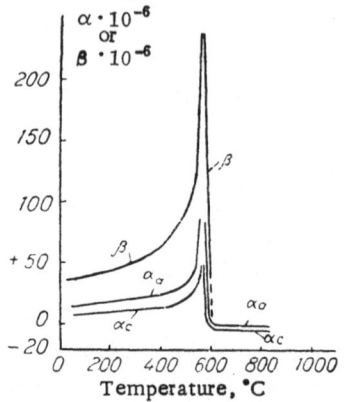

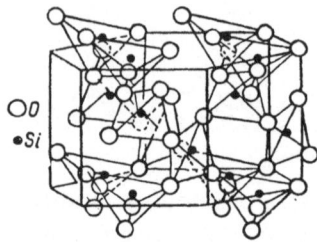

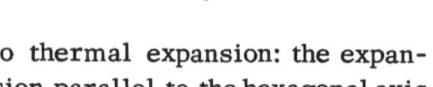

Fig. 43. Crystal lattice of β-quartz (low-temperature).

Fig. 44. Thermal expansion coefficient of quartz, according to Jay (1933), in relation to temperature: α_a) linear along the a axis; α_c) the same along the c axis; β) volume.

to thermal expansion: the expansion parallel to the hexagonal axis is a factor of almost 1.6 less than that normal to this axis. The reason for this can be seen by examining the structure of quartz, which has a considerably lower density of packing of the silicon and oxygen atoms along the hexagonal axis than in planes normal to it. After the conversion into the α-form (high-temperature), the expansion anisotropy decreases and contraction begins first in the direction of the hexagonal axis [apparently reduction of the pitch of the spiral chains of (SiO₄) tetrahedra] and then in other directions. The reasons for the latter are still not clear.

The change in the heat capacity of quartz with temperature is shown in Fig. 55. The sharp increase in the heat capacity of β-quartz close to the conversion temperature, like the increase in the thermal expansion coefficient (Fig. 44), indicates "preparation" for reorganization of the structure.

Sosman [7] pointed out that there is a secondary structure in quartz, which causes the presence of very fine channels along the hexagonal axis in which there may be foreign substances. The time dependence of the polarization of quartz in an electric field indicates the inhomogeneity of quartz. Shishakov [301] drew attention to the fact that the pyknometric specific gravity of quartz (2.649) is almost 1.5% less than that calculated from the lattice parameters (2.688). This was explained by the presence of a mosaic structure with a fissure thickness of 6.5 A. An electron microscope investigation of quartz [302] saturated with water at 360° and a pressure of up to 300 atm showed the presence of a mosaic structure. The aggregates had a thickness of about 0.9 μ and a length of 9–50 μ. The presence of a mosaic structure explains the capacity of quartz to hold inclusions (for example, in amethyst) and the penetration of cations through it under applied potential differences.

The valence oscillations in quartz, which are restrained below 25°, begin to be excited with an increase in temperature [303] and at 130° all the lattice points have at least one excited bond and 5% have an excited pair of bonds. At 430° all the lattice points of silica have excited pairs of valence bonds and 15% have three or four. It is possible that rotation of the oxygen atoms in orbits of small diameter begins here as with cristobalite. The (SiO_4) tetrahedra become regular and at 573° β-quartz changes into the α-form (high-temperature), whose structure is very close to that of β-quartz [297–299] but differs from it in that all the (SiO_4) tetrahedra are regular, which leads to a certain increase in the symmetry of the crystal lattice. The (SiO_4) tetrahedra in α-quartz are inclined to each other at an angle of 150° (see Fig. 49). This conversion was first described by Le Chatelier [304]. The modification change of quartz is accompanied by the formation of fissures and twinning [306]. Tsinzerling [305] showed that quartz that had been purified electrically at high temperature lacked the capacity to change its modification or twinning, but the reasons for this have not been established as yet.

The phase change in quartz is preceded by the formation of defects in the lattice. However, it was found [697] that the amount of defects in the β-quartz lattice at 548° was only 0.0025%. It was shown experimentally [698] that the reflection of light occurs at the β-α-quartz phase boundaries. Close to the phase change in quartz, a dark band appears in monocrystals, covering a region of 0.1°. In this band the intensity of light scattering is $1.4 \cdot 10^4$ greater and the depolarization of the scattered light is a factor of two less than normal [699]. The dark band separates the regions of β- and α-quartz, which disproves the hypothesis of Steinwehr [700] on the presence of intermediate forms of quartz in the β-α conversion. The phase changes of silica and the change in the nature of the bond at this point were studied by Badaluco [704].

Centers causing the smoky color of quartz are produced by the ejection of electrons from the crystal lattice and their fixation at various levels due to impurities. The latter are partially removed by heating to 251° [307], but sometimes remain even with heating to a higher temperature.

Accurate dimensions of the elementary cell of quartz, according to Keith [308], are presented in Table 16.

Table 16

Dimensions of the Elementary Cell of Quartz

Origin of quartz	Preparation temperature, °C	Dimensions of elementary cell, A	
		a	c
Natural Brazilian	—	$4{,}91248_8$	$5{,}40422_7$
Synthetic	388	$4{,}91263_6$	$5{,}40439_6$
»	369	$4{,}91274_2$	$5{,}40442_5$
»	292	$4{,}91312_3$	$5{,}40471_3$

The table shows that there are changes in the size of the edges of the quartz cell in relation to the synthesis temperature. This is apparently connected with the presence of different amounts of impurity in it.

Keith and Tuttle [309] used thermal analysis to study the conversion of 250 quartz samples. The differences in the conversion temperature for samples of natural quartz did not exceed 38°, while for artificial quartz they were up to 160°. More than 95% of the natural quartz samples showed deviations from the conversion temperature of up to 2.5°. Variations in the conversion temperature are probably connected with the formation of solid solutions (the presence of ions other than those of silicon and oxygen in the quartz lattice) and this was proved by spectral and x-ray investigations. In the presence of germanium, synthetic quartz gave a conversion temperature of 40° above, and in the presence of lithium and aluminum, 120° below the normal (573.2°). Usually the amount of impurities depends on the crystallization temperature of the quartz. Therefore, in a series of cases, the temperature of the β-α conversion of quartz may serve as an indirect indication of the crystallization temperature of quartz, but only in rocks which do not differ too much in composition. The concentration of the solid solutions causing the change in the conversion temperature of quartz is not known, but it must be very low.

The β-α conversion temperature hardly depends on the grain size of the quartz (up to 0.9°). The conversion begins in all directions, probably from the centers of local deformations, which may be, for example, the surroundings of inclusions. Most frequently of all, the conversion of quartz becomes noticeable at the grain surface [643]. No thermal effect was observed in the conversion of flint and novaculite, although x-ray pattern data show that they consist of quartz grains. Crushing novaculite to the size of monocrystals made it possible to detect the thermal effect of the β-α conversion of quartz. The absence of this effect with uncrushed samples is explained by displacement phenomena, leading to a "diffuse" conversion temperature and liquidation of the "sharp" temperature peak on the thermogram.

We should remember that Myuller [303] reported the "diffuseness" of the β-α quartz conversion for monocrystalline samples also.

Depending on the origin, natural quartz shows measurable variations in the relative content of the oxygen isotopes O^{18} and O^{16} and this must also have an effect on the β-α conversion temperature. However, this problem has not been studied at all.

The presence of several peaks on the thermograms of quartz is caused by its zonal structures. In itself, the conversion of quartz does not depend on solid or gaseous inclusions in it, for example, of the amethyst type, but this sometimes leads to a nonisothermal conversion [310].

The conclusions of Keith and Tuttle [309] were confirmed by the investigations of Sabatier and Wyart [311], who showed that for synthetic quartz the presence of the sodium ion raised the β-α conversion temperature of quartz by 7° and was accompanied by a decrease in a_0 by 0.0015 kX, while the size of the cell edge c_0 remained unchanged.

In connection with the solution of geological problems, the effect of pressure on the β-α conversion temperature of quartz was studied [312, 313] and in the latter work, the pressure was raised to 10,000 atm: this study gave the following results:

Excess pressure, bars	Conversion temperature, °C
0	572.3
1,000	599
2,000	626
3,000	653
5,000	704
7,000	751
10,000	815

The initial pressure effect equals $29 \cdot 10^{-3}$ deg/bar and at 10,000 bars, it is about $20 \cdot 10^{-3}$ deg/bar, i.e. differing little from the initial.

This relation makes it possible to use the Clapeyron equation to determine the thermal effect of the β-α quartz conversion and this was found to be 4.2 cal/g, while according to thermal analysis it is 1.75 cal/g, which is apparently somewhat low.

The increase in the temperature of the β-α quartz conversion in relation to pressure is given by the equation

$$\Delta T = -1.6 + 2.871 \cdot 10^{-2}P - 4.284 \cdot 10^{-7}P^2,$$

where P is the pressure in bars (1 bar = 1.0197 kg/cm^2). From the equation given it follows that the maximum conversion temperature of quartz is 1050° and at a pressure of 33,500 bars, but this still requires experimental verification.

Sinel'nikov [314], working with an adiabatic vacuum calorimeter which he constructed himself, found that in contradiction to reports, the β-α quartz conversion is a phase change of the second order without the isometric absorption of heat. The anomalous increase in the heat capacity of quartz with a rise in temperature shows that the phase change begins at about 533° and is complete at 577°. The heat consumed in the quartz conversion in this temperature range is 9.2 cal/g according to Sinel'nikov. These ideas are in accordance with the mechanism presented above for the β-α conversion of quartz, though they are not comprehensive. An examination of the accumulated data shows that there is "preparation" for the β-α conversion of quartz at comparatively low temperatures, but it is completed suddenly at a definite temperature.

Birks and Schulman [378] established that a short duration firing of amorphous silicic acid with the addition of carbonates of Group II elements and manganese at 1200° led to the formation of quartz and not tridymite as should be the case according to classical ideas. In the presence of other additives, amorphous silicic acid changes into cristobalite and not tridymite at 1200°. These data indicate the need for re-examining classical ideas on the temperature region in which quartz is stable.

By studying changes in the elastic characteristics of quartz during piezoelectric oscillations in relation to temperature, Osterberg and Cookson [315] found that quartz changed into a new form at 847°, which was not piezoelectric. Since this conversion was rapid, these investigators put forward the hypothesis that a new modification, γ-quartz, was thus formed.

Analogous experiments then led Osterberg [316] (see also [17]) to the conclusion that at −183.5° (89.7°K) normal piezoelectric quartz changed into a new, hexagonal, but nonpiezoelectric modification, δ-quartz.

However, the more specific, accurate investigations of Pavlovich and Pepinskii [317] of the electroresonance, dielectric constant, and volume changes of quartz and its oscillograms and x-ray patterns by Weissenberg's method showed the error of Osterberger's conclusions. It was established that there are no quartz conversions in the temperature range from 4.2 to 846°K (from −269 to 573°). The physical reason for this is that the valence oscillations are practically completely suppressed even at room temperature in quartz and other modifications of silica [303]. Therefore the "rigid" crystal lattice of silica cannot undergo spontaneous rearrangement which also requires a certain change in the length of the valence bonds.

The change in the elastic properties of quartz proceeds in jumps at 370 and 510° [678]. The elastic constants of quartz pass through a minimum at the conversion point [701].

With prolonged heating above 867° [401] in the presence of mineralizers, quartz is converted into tridymite and above 1027° in the absence of mineralizers [352], into cristobalite. Its metastable melting point at normal pressure is 1610°, while in the presence of water under a pressure of 1400 atm, quartz melts at about 1100° [318].

Eustachiv and Greenwald [319] showed that quartz recrystallizes when heated while the logarithm of the recrystallization time depends linearly on the reciprocal absolute temperature of the process. It is still not possible to obtain large monocrystals of quartz required for practical use by this method. However, quartz monocrystals weighing more than 450 g were obtained under hydrothermal conditions at 380–400° and 1050 atm after 80 days and this corresponds to a growth of more than 2.5 mm per day.

In other work, quartz monocrystals weighing more than 400 g were obtained under hydrothermal conditions after 60 days. Besides being of great technical importance, these experiments disproved the opinion of geologists that quartz monocrystals must grow slowly in nature.[1]

In connection with the great importance of quartz and cristobalite, their heats of formation were recently determined more accurately (on preparations of 99.9% purity) [322] and these are given in Table 15.

Investigation of the infrared spectra of quartz showed [323] that low temperature β-quartz has rings (at 1055 cm^{-1}) indicating stretching of the valences $\overleftrightarrow{Si}$–$\overleftarrow{O}$–$\overleftrightarrow{Si}$.

The optical properties of quartz are given in Table 17.

The data presented show that the structure and properties of quartz are not as simple and unequivocal as normally seems to be the case from a study of the optical properties of quartz with moderate accuracy and the use of its elastic properties. Up to now, quartz has not been obtained in the absence of a melt or volatile components. Under these conditions its conversion into cristobalite is essentially monotropic. Even with prolonged heating of high-silica, high-density Dinas, used in the crown of a Martin furnace (several months), according to the data of Kainarskii and Karyakin [324] there is no reverse change of cristobalite into quartz at temperatures which correspond to its classical stability range. This, together with the ever-increasing amount of new data on the presence of impurities in quartz and on the genesis of quartz compels a reexamination of the problem of its position in the phase diagram of silica. There should be a special investigation of the question of whether or not the structure of quartz is generally stabilized by

---

[1]The problem of the hydrothermal synthesis of different modifications of SiO$_2$ is of separate interest and therefore is not considered here.

Optical Properties and Specific Gravity of Various Forms of Silica

Form	Specific gravity, g/cm³	Crystal system	Optical properties						Average refractive index	Specific refraction energy
			Ng	Nm	Np	Ng−Np	2V	Optical character		
Fibrous modification (silica)	1.97	Rhombic	—	(1.408)	—	—	—	—	(1.408)	(0.207)
β-Quartz (low-temperature)	2.65	Hexagonal	Ne 1.553	No 1.544	—	0.009	—	Positive	1.547	0.207
α-Quartz (high-temperature)	2.52	"	Ne 1.540	No 1.532	—	0.008	—	»	1.535	0.211
γ-Tridymite (low-temperature)	2.31	Rhombic	1.473	1.469	1.469	0.004	43°	»	1.470	0.204
β¹-Tridymite (117−163°)	(2.29)	Hexagonal	—	1.475	—	—	—	—	1.175	(0.207)
β²-Tridymite (163−210°)	—	—	—	—	—	—	—	—	—	—
β³-Tridymite (210−475°)	—	—	—	—	—	—	—	—	—	—
α-Tridymite (high-temperature)	2.23	Hexagonal	—	(1.477)	—	—	—	—	(1.477)	(0.207)
β-Cristobalite (low-temperature)	2.34	Tetragonal	—	No 1.487	Ne 1.484	0.003	—	Negative	1.486	0.208
α-Cristobalite (high-temperature)	2.22	Cubic	—	No 1.466	—	—	—	—	1.466	0.210
Keatite	2.50	Tetragonal	—	No 1.522	Ne 1.513	0.009	—	Negative	1.519	0.208
Coesite	3.01	Monoclinic	1.604	—	1.599	0.005	—	Positive	1.602	0.200
Lechatelierite (siliceous glass at normal pressure)	2.20		—	1.458	—	—	—	—	1.458	0.208
Supraplezo glass	—	—	—	—	—	—	—	—	—	—
Condensed glass	2.61	—	—	(1.540)	—	—	—	—	(1.540)	(0.207)

Note: The average specific refraction energy of silica $K = 0.207$ and the values in brackets are calculated.

foreign atoms (ions) present in the lattice in small amounts, as was recently found for γ-Al_2O_3. In addition to metal cations, hydrogen, hydroxyl, and water must also be considered. There is no analogy between SiO_2 and GeO_2. The most thoroughly studied modification of GeO_2 crystallizes in a rutile form, while the poorly studied "soluble" form of it crystallizes like quartz [402]. However the individuality of this form of GeO_2 raises doubts [56] in precisely the same direction as for quartz.

When α-quartz is heated to about 1000°, it changes into α-cristobalite (see Fig. 48) and this is accompanied by "straightening" of the (SiO_4) tetrahedra [7, 292, 324, 339–341]. In the absence of a melt or volatile substances, the conversion of quartz into cristobalite is irreversible.

Kainarskii and Degtyareva [342] showed that regardless of temperature, cristobalite was always obtained at first in the crystallization of quartz glass. It is known that amorphous silica changes into cristobalite when heated. This modification of silica is also formed during crystallization from the gas phase [675]. The effect of additives on the conversion of silica and the existing rules on this were studied by Kainarskii and Orlova [673].

The grain size was found to have some effect on the rate of conversion of quartz into cristobalite [341]. Quartz grains 4–5 mm in size were completely converted into cristobalite at 1500° in two hours at 1500° and fine grains (up to 4 μ) were converted even at 1300°. The kinetics of quartz conversion under the conditions for the production of Dinas (addition of up to 2% of CaO to quartzite) were investigated by Avgustinik and Kurdevanidze [344], who established that the rate constant of quartz conversion, K, depends on the grain diameter, D, and the absolute temperature, T, according to the equation

$$K = \exp\left(\frac{a}{D^{0.064}} - \frac{b \cdot 10^4}{T} \right),$$

where a and b are constants. This formula has recently been criticized [706]. In the light of new data, the ideas on the existence of two sorts of cristobalite molecules put forward by Smith lack foundation.

The effect of mineralizers on the conversion of quartz was studied recently by Grimshaw [371].

On the basis of structural data and symmetry, it is usually considered that the angle between Si–O–Si bonds of two adjacent

(SiO$_4$) tetrahedra in cristobalite equals 180° and on this basis it was stated that the bonds in cristobalite have an ionic character. However, it is possible [327] that the angle deviates somewhat from 180°. The high symmetry of cristobalite is caused by rotation of oxygen atoms in small orbits and the bonds have a homeopolar character to some extent.

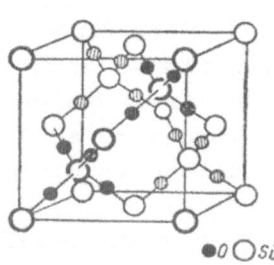

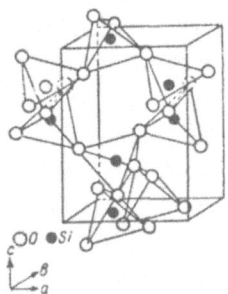

Fig. 45. Crystal lattice of α-cristobalite (high-temperature).

Fig. 46. Crystal lattice of β-cristobalite (low-temperature).

The structure of cubic α-cristobalite was studied by Wyckoff [328], Barth [329], and others. As a result of this, it was established that the silicon atoms in the crystal lattice of α-cristobalite (high-temperature) are arranged as in diamond (Fig. 45). The characteristics of the cristobalite lattice are presented in Table 2. The latest data [330] confirmed that α-cristobalite belongs to the cubic system, as was found by Wyckoff and contrary to the opinion of Barth. If there are some deviations from the cubic system, then they are quite insignificant in this case.

At 218 ± 2° α-cristobalite changes into the low-temperature β-form [330]. The transition occurs at the same temperature with heating. Hysteresis in the β-α conversion of cristobalite is caused by the presence of impurities, which are frequently found in it [331]. This conversion occurs due to the rearrangement of some elements of the crystal lattice without an essential change in the bonds as in the β-α conversions of quartz. At this, the rotation of the oxygen atoms in orbits of 0.4 A radius is stopped. The bond angle in the low-temperature form, β-cristobalite, is about 150° [332]. Gibbs [333] also considered that the β-α conversion of cristobalite is connected with rotation in the (SiO$_4$) tetrahedra.

α-Cristobalite contains chains of (SiO_4) tetrahedra aligned along a second order axis. These chains are somewhat deformed in β-cristobalite, whose structure is tetragonal [330, 336], as was also established previously [332], contrary to Barth [334], who considered it rhombic, or Belyankin and Ivanov [335], who even assigned it to the triclinic system. The structure of β-cristobalite (low-temperature) is presented in Fig. 46.

Melanophlogite, which has a refractive index of 1.461 and a specific gravity of about 2.04, is essentially a cristobalite [37] and not a quartz as Ormont stated [56]. In exactly the same way, the fibrous minerals lussatite and lussatine are cristobalites [337].

As Belyankin showed [338], metastable cristobalite which is formed from quartz is practically isotropic even at room temperature.

Twinning occurs during the conversion of α-cristobalite monocrystals into the β-form [330]. The properties of cristobalite at very low temperatures have not been studied as yet. On the basis of the ideas developed by Myuller [303], there are no grounds for considering that cristobalite may have further conversions at temperatures below 0°. This is confirmed by the results of studying the infrared spectra of cristobalite [323] in the temperature range 4–880°K and also the change in the heat capacity with temperature, beginning at 5°K [705].

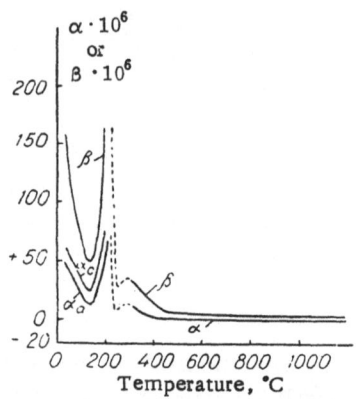

Fig. 47. Thermal expansion coefficients of cristobalite according to Johnson and Andrews (1955) in relation to temperature: α_a) linear along the a axis; α_c) the same along the c axis; β) volume.

The thermal expansion of cristobalite was studied by Johnson and Andrews [330] with the aid of a high-temperature x-ray camera. The changes in the expansion coefficients of cristobalite, based on these investigations, are shown in Fig. 47. The details of the changes in the expansion coefficients, for example, the clearly expressed minimum at about 120° and also the small maximum at 300°, cannot be explained as yet. As for quartz, there is a sharp increase in the expansion coefficient at temperatures close to the conversion. The conversion of β-cristobalite into the α-form is accompanied by a volume increase of 3.7%. Büssem [345] studied the thermal expansion of cristobalite obtained in the presence of fluxes. He found that the cristobalite expansion curve has a maximum at about 1000°; this disagrees with the data of Johnson and Andrews. Trömel and Obst [645] found that the conversion temperature of cristobalite was 226°. Flörke [646] showed that this temperature depends on the standard of crystallization of the cristobalite and for very good crystals, it is 270–275°.

Application of the Clapeyron equation (see initial data in Table 18) shows that the effect of pressure on the conversion temperature of cristobalite must be considerably greater (about 2.3 times) than for quartz conversion. This dependence has not been studied experimentally as yet.

Fenner [292] added Na_2WO_4 as a flux to facilitate silica phase conversions and established that cristobalite is unstable below 1470° and changes into a new modification of silica, tridymite. This conclusion was incorporated in all phase diagrams of systems containing silica, for the most part without further experimental proof. However, Peyronel [346] shook the generally accepted conclusion of Fenner to the roots by showing that cristobalite was formed quantitatively at 850° in the presence of small amounts (1 molecular part of X_2O_5 + 100 molecular parts of Na_2WO_4 + 50 molecular parts of SiO_2) of oxides of Group V elements (P_2O_5, V_2O_5, or As_2O_5). This obviously raised doubts on at least the lower limit of the stability region of cristobalite, established by Fenner's method. We consider that Peyronel's experiment is of much more importance and this, together with the results of other investigations (see below), indicates the need for new experiments to establish the position of tridymite on the phase diagram of silica. In this connection, we should also remember the experiments of Khrushchev [349], who established that cristobalite crystallized at comparatively low temperatures in the presence of water.

Kainarskii and Degtyareva [347] studied the action of a large number of additives on the conversion of silica and came to the conclusion that this was apparently determined by the structure (closest order) of the melt formed. This is in accordance with the results of Peyronel's experiments.

Sosman [348] established that cristobalite melts at 1723°. The latest determinations gave its melting point as 1728°. The melting point of cristobalite is depressed by more than 250° in the presence of water vapor under a pressure of 300 atm [318].

The temperature dependence of the vapor pressure of various modifications of silica has not been studied at all. Kamentsev [150] presented graphical data on the vapor pressure of silica (P, mm Hg) in relation to absolute temperature, T, which may be expressed by the equation

$$\lg P_{SiO_2} = 14.03 - 26\,200 \cdot T^{-1} \text{ [mm Hg].}$$

Elyutin [544] presented the following equation for this purpose:

$$\lg P_{SiO_2} = 7.65 - 19\,130 \cdot T^{-1} \text{ [atm]}$$

It hence follows that the vapor pressure of silica reaches 1 atm at approximately 2080°. It is 0.012 atm at the melting point of cristobalite and at the transition temperature of quartz into tridymite (if it is considered that this equation is accurate at such temper-

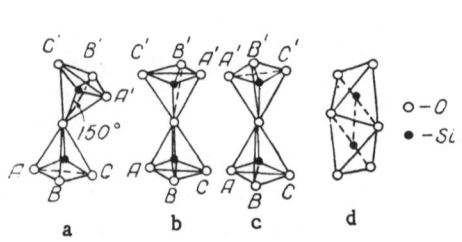

Fig. 48. Different methods of joining (SiO₄) tetrahedra in SiO₂ modifications: a) α-quartz; b) α-cristobalite; c) α-tridymite; d) fibrous modification of silica-W.

Fig. 49. Crystal lattice of α-tridymite (high-temperature).

atures), it is only of the order of 10^{-9} mm Hg, i.e., a quite insignificant value. Despite all the possible errors in estimating the value of the vapor pressure of silica at these temperatures, it is obvious that the vapor pressures of different modifications of silica differ so little that they cannot have any effect either on the stability of the modifications or on the kinetics of their conversions. Therefore the usually cited phase diagram of silica on $P-T$ coordinates in accordance with Fenner's treatment with a schematic indication of differences in the vapor pressure of silica modifications should be abandoned as it has no real meaning. Other diagrams presented below should be used instead of it.

Table 18

Characteristics of Silica Phase Conversions

Conversion	Characteristics of conversion	Temperature, °K	ΔH, cal/mole	ΔV, cm³/mole	$\frac{dP}{dT}$ at conversion point, kg/cm²·deg	Angle of deviation from vertical, α, deg	tg$^{\alpha}$
β-Quartz → α-quartz	Rapid	846	250	0,295	64,7	0,97	0,0155
α-Quartz → α-tridymite	Slow	1140	120	3,489	1,29	37,79	0,775
α-Tridymite → α-cristobalite	»	1743	50	—0,1	—12,249	—4,69	—0,080
α-Cristobalite → melt . . .	»	2001	1835	0,028	1399,2	0,04	0,0007
β-Cristobalite → α-cristobalite	Rapid	491	310	0,970	27,8	2,05	0,036
γ-Tridymite→β₁-tridymite. . .	»	390	70	0,053	142,3	0,39	0,007
β₁-Tridymite→ β₂-tridymite . .	»	436	40	0,619	25,3	0,22	0,040
β₂-Tridymite→ α-tridymite . .	»	483	45	—0,484	—8,21	—0,69	—0,012

Note: From results of thermal analysis [693], the heat of the β-α conversion of cristobalite was determined as 192 cal/mole, which is apparently somewhat low.

When one (SiO$_4$) tetrahedron is rotated through 180° with respect to another (Fig. 48), a new modification of silica, namely tridymite, is produced and this exists in several forms.

The change from the structure of quartz to that of tridymite requires not only straightening, but also the given rotation of the (SiO$_4$) tetrahedra, which involves "breaking" of $^3/_4$ of the Si–O bonds. These processes undoubtedly require a larger consumption of energy than just for straightening the tetrahedra. Consequently quartz changes into cristobalite much more readily than into tridymite and therefore cristobalite is the intermediate phase in the conversion of quartz into tridymite. Breakage of the Si–O bonds in the rotation of the (SiO$_4$) tetrahedra during the conversion of the cristobalite structure to that of tridymite also requires a large activation energy and is difficult to achieve.

The structures of the low- (γ) and high-temperature (α) forms of tridymite were investigated by Gibbs [350] and also partially by Schiebold [351]. The results of these investigations are given in Table 2 and the structure of α-tridymite, in Fig. 49. The optical properties of tridymite are presented in Table 17.

As in quartz, the (SiO$_4$) tetrahedra in the low temperature form of tridymite are somewhat deformed. Like the cristobalite lattice, the tridymite lattice is built up from flat, hexagonal (Si$_2$O$_5$) meshes. In tridymite the structure is repeated every third layer and in cristobalite, every fourth. The peculiarities in the structure of tridymite are caused by the presence of tubes penetrating the whole structure, whereas closed voids are formed in cristobalite.

Fenner established that tridymite has conversion points at 117 and 163°.

As in cristobalite, the presence of phase changes in tridymite is connected with rotation of the oxygen atoms, which are in two groups here [352]. It is probable that one group of oxygen atoms stops rotating at 163° and the other at 117°. The lower temperature of these conversions in cristobalite and tridymite than in quartz is apparently explained by the looseness of their structures. These ideas, however, do not explain the new conversions in tridymite, which occur at 210–225° [352, 353] and 475° [353] and although they give a considerably smaller effect on the expansion curve (Fig. 50) than the conversions at 117 and 163°, nonetheless, they cannot be considered as phase changes of the second order [354]. The nature of these conversions has not been elucidated as yet. At the moment we cannot completely exclude the possibility that the conversion at 210–225° was caused by the presence of small amounts of cristobalite [643] and that at 475°, by small amounts of quartz. Due to twinning and the fineness of tridymite crystals, the thermal expansion along its crystallographic axes has not been measured up to the present. This hampers an examination of expansion and phase conversions of tridymite. Figure 50 shows that the thermal expansion coefficient increases sharply toward the conversion points at 117 and 163° in the case of tridymite also. This coefficient has a negative value

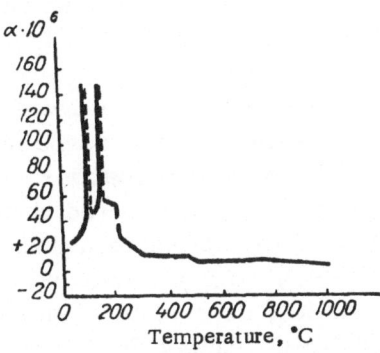

Fig. 50. Average linear thermal expansion coefficient of tridymite in relation to temperature according to Austin (1954).

in tridymite above 900°. The thermal expansion of polycrystalline tridymite, obtained by devitrification of quartz glass with small additions of alkali, is characterized by the following data (according to Kainarskii and Degtyareva [VNIIO, 1955]):

Temperature, °C	Expansion, %	Temperature, °C	Expansion, %
100	0.21	700	1.20
200	0.80	800	1.22
300	0.96	900	1.24
400	1.11	1000	1.25
500	1.15	1100	1.22
600	1.16	1200	1.20

The linear thermal expansion coefficients of this tridymite have the following values:

Temperature range, °C	Average value of $\alpha \cdot 10^6$
100–1200	9.0
100–400	90.0
400–1200	1.1

Obviously, the low thermal expansion of tridymite is of practical interest only under conditions when the temperature is above 400° and this is of importance in the use of tridymite as a refractory material. The metastable melting point of tridymite equals 1703° [352]. Under hydrothermal conditions, tridymite melts directly into a hydromelt [318].

Sinel'nikov [340] showed that fineness of powdering of quartz is of secondary importance in the conversion into tridymite.

The optical properties of natural tridymite are variable [37]. Lukesh [355] considered that tridymite contains traces of foreign ions so that its composition reduces to the formula $NaCaAl_3Si_{15}O_{36}$ (79% SiO_2 – A.B.). Barth and Kvalheim [356] found 5.2% of nepheline, $NaAlSiO_4$, as a solid solution in tridymite and this corresponds to a 97% SiO_2 content. In this case there is only one conversion at 135°. Flörke [357] showed that the conversion temperatures of cristobalite and especially that of tridymite depend on their preparation conditions. Apparently, the presence of a small amount of alkali in the crystal lattice is generally necessary for the formation of tridymite. Sosman [354] proposed the possibility of considering tridymite as a heterogeneous substance. The proposal of Tuttle and Insley on the existence of two "sorts" of tridymite with conversions at 117 and 163°, respectively, is not confirmed by the latest investigations. The so-called chrystensite has been found to be normal tridymite by x-ray investigation [358]. The existence of pseudomorphoses of cristobalite, as tridymite, under conditions when the primary separation of the latter is probable [359, 360] in some rocks shows that the interrelation of these modifications of silica requires further study.

Ruddlesden [361] examined the electrostatic energy of the crystal lattices of tridymite and cristobalite in connection with the crystallization (devitrification) of quartz glass.

The structure of cristobalite has cubic closest packing of the (Si_2O_5) layers and tridymite, hexagonal closest packing. The Si-O, O-O, and Si-Si distances in them are the same for nearest "neighbors," but differ somewhat for the more distant atoms. Calculations of the bond energies for the two types of lattice showed that there is an energy minimum with the cristobalite structure and a maximum with the tridymite structure. Therefore the structure of cristobalite is found to be more stable and, apart from other reasons, is obtained more readily in the crystallization of silica glass.

The existence of fibrous varieties of chalcedony [362] suggested the possibility of the existence of a special modification of silica with a structure close to that of silicon chalcogenides. A fibrous mass of silica is formed by volcanic activity, in cold parts of blast furnaces and also in the preparation of SiO [254, 269]. The new form of silica was recently obtained by Weiss [363] by treatment of finely ground quartz and silicon, mixed in the ratio required for the preparation of SiO, in vacuum at 1250–1500°. With the exclusion of air, only SiO is obtained in this case. When pure oxygen was introduced into the reaction sphere (pressure of about 0.001 mm Hg), a fibrous variety of SiO_2 was obtained and this condensed on a condenser tube.

$$2SiO_{gas} + O_2 = 2SiO_2.$$

Its fibers were very flexible and strong. The preparations obtained contained 97.2–98.9% SiO_2, 0–0.9% Si, 0–2.9% SiO, 0.1–0.6% Al_2O_3, 0.2% N, and traces of Fe. The properties of the new silica modifications are presented in Table 2. Sosman [354] proposed the name of silica-W for this form of silica.

Silica-W fused when heated to 1420°, but the fibers which did not melt changed into cristobalite. When it was kept at temperatures of 200–800° for a day, tridymite was formed from it and this occurred the more readily the greater the Al_2O_3 content of the original preparation. An analogous conversion was observed when silica-W was kept at room temperature for 20 months. In all the cases the fibrous habit was retained. When treated with water, silica-W changed into amorphous silicic acid. This form of silica was also created during the reduction of silica or silicates to SiO or silicon at 1200°, for example, with carbon or aluminum in a graphite crucible (at 1380°).

The characteristic of the silica-W structure is that the linking of the (SiO_4) tetrahedra is not through the apexes, but the edges (see Fig. 49) like the (SiS_4) tetrahedra in the structure of SiS_2 (see below). Each silicon atom in silica-W is in the middle of a deformed tetrahedron of (SiO_4). This results in a one-dimensional, infinite molecular chain:

The Si–O distance in this structure equals 1.87 A, which is considerably greater than that in low-temperature quartz (1.61–1.62 A) or cristobalite (1.59 A) and corresponds almost exactly to the calculated distance for the homeopolar Si–O bond (1.83 A). The Si–Si distance in silica-W equals 2.28 A. The shortest separation of oxygen atoms in nearest molecular chains is 3.51 A.

According to Pauling's rules, the presence of (SiO_4) tetrahedra linked by edges makes this structure unstable.

Silica-W is the least dense form of SiO_2.

If silica-W is left in moist air, its structure is destroyed and its fibers become amorphous. During this process, 0.01–0.08 molecules of H_2O are absorbed per molecule of SiO_2 and metasilicide acid is formed and this is accompanied by cleavage of the bond between two oxygen atoms in the chain of (SiO_4) tetrahedra by the scheme

as a result of which the distances between the silicon atoms in the chain increase. Therefore the fibers must extend or bend. With the absorption of water vapor, screw spirals are formed from the (SiO_4) tetrahedra. If the water vapor content of the air is low, silica-W changes into a glass at very high temperatures.

Apparently, finely fibrous varieties of chalcedony, consisting of quartz [362] or cristobalite, are produced from the fibrous form of silica. Crystals of quartz [325, 364] and cristobalite [365] in chalcedony fibers are oriented and may be pseudomorphs of silica-W.

A study of the effect of sodium on the crystallization of amorphous silica under hydrothermal conditions led to the discovery of a new modification of silica [366], for which Sosman [367] proposed the name "keatite" in honor of its discoverer. The formation of a new silica phase under these conditions may have been surmised from the experiments of Endell [368]. Sosman's co-worker Keat obtained the new modification of silica at 380–585° and a pressure of 350–1265 kg/cm^2 from silica gel in the presence of about 1 ml of 0.01 N NaOH to 1.5 g of SiO_2. Cristobalite was formed with a smaller amount of sodium hydroxide, and quartz, with a greater amount. Keatite may give polycrystalline formations. Spectral and chemical analyses showed that the new phase was pure silica and it dissolved completely in cold hydrofluoric acid. When keatite was heated to 1100° and kept at this temperature for 37 hours, only insignificant conversion to cristobalite was observed. Conversion to cristobalite was complete at 1620° in three hours. The usual changes of SiO_2 were observed on heating in the presence of sodium tungstate. No conversion of keatite was noted on thermograms up to 1100°.

The properties of keatite are presented in Tables 2 and 17. A study of the change in dimensions of the elementary cell showed that keatite expands parallel to the c axis up to 550° and contracts up to 250° and then expands parallel to the a axis. The over-all effect of thermal expansion for keatite aggregates is shown in Fig. 51. Judging by x-ray powder patterns, the structure of keatite is similar to that of β-spodumene, β-$LiAlSi_2O_6$, which is apparently derived from this new form of silica. The μ-

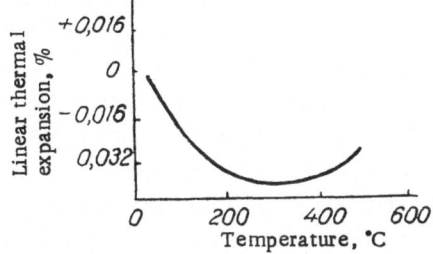

Fig. 51. Temperature relation of the linear thermal expansion of polycrystalline samples of keatite from 25°, calculated from the change in dimensions of the elementary cell, according to Keat (1954).

phase of cordierite, $Mg_2Al_4Si_5O_{18}$, has a similar type of structure. The similarity of these minerals is also indicated by the character of their thermal expansion.

The structure of keatite is characterized by the presence of double spirals of (SiO_4) tetrahedra. Here, as in quartz, there are two types of (SiO_4) tetrahedra, differing in position; consequently, keatite must be dextro- and levorotatory. The presence of twins is assumed. As yet there has been no detailed investigation of the structure of keatite. Keatite is converted into low-temperature quartz [354].

It is possible that keatite is also formed by careful heating of silicic acid to 1000–1300°.

While investigating conditions for the preparation of some minerals under high pressure, Coes [369] obtained a new form of silica, for which Sosman [367] proposed the name "coesite." The best formation of coesite was observed in a mixture of equal amounts of dry sodium metasilicate and diammonium phosphate at 750° under a pressure of 35,000 atm for 15 hours. The new form of silica appeared as tabletlike, hexagonal crystals up to $50\,\mu$ in size. The diammonium phosphate may be replaced by boric acid, ammonium chloride, or vanadate and also potassium fluoborate. Potassium silicate may be used instead of sodium silicate.

Coesite is also obtained under the conditions mentioned from silicon and silver carbonate, which indicates that the presence of hydrogen is not essential to its formation.

The new form of silica has a very high specific gravity (see Table 2). It is the most dense modification of SiO_2. When heated to 1700°, coesite changes into cristobalite and silica glass. The considerable stability of coesite has made it impossible as yet to determine its field of stability on the $P-T$ diagram. This modification is not formed at pressures below 35,000 atm. Its best formation was observed at temperatures of 500–800°, but even below 500°, silica is slowly converted to coesite. Quartz is obtained at pressures below 35,000 atm and higher temperatures.

Coesite is stable at a pressure satisfying the relation [703]

$$P \geqslant 22.5\,T + 9500 \text{ bars.}$$

The heat of conversion of quartz into coesite equals -225 ± 150 cal/mole and the entropy of coesite at a pressure of 1 bar and a temperature of 250° is 8.6 ± 0.7 cal/deg·mole. Under conditions in the earth's crust, coesite should be stable at depths of 60–100 km.

Coesite dissolves very slightly in hydrofluoric acid, but dissolves rapidly in molten ammonium difluoride.

The properties of coesite were investigated by Coes [369] and Ramsdell [370]. The results of their investigations are presented in Tables 2 and 17. Ramsdell reported a very great disagreement between the calculated value of z for the elementary cell of coesite and what it should be according to the space group. The reason for this has not been established as yet. The crystal structure of coesite and also its properties (for example, thermal expansion) have not been investigated in detail as yet. The existence of this modification in rocks has not been proved yet.

During the mechanical treatment of quartz (grinding and polishing) a change occurs in it, namely decrystallization [354], and this was observed by Ray even in 1923. This phenomenon does not lead to the formation of an amorphous phase, but is accompanied by a fall in specific gravity and the heat of solution in hydrofluoric acid. However, a new modification of silica is not formed during this process. These changes require further investigation.

An exceptionally important form of silica from the practical point of view, which is of interest in science, is its glass, which is usually known as quartz glass. The structure of quartz glass has been studied by many investigators [379, 380, 382–386, 674], and its properties have also been described [7, 387]. Ignoring the special problem of the structure of glasses, we will only state that quartz glass is considered to be made up of (SiO_4) tetrahedra attached by the vertices.

A series of data (a study of the effect of temperature on the refractive index [388], electroconductivity [389], and scattering of x-rays [380, 381]) show that quartz glass contains formations which most investigators consider as cristobalite and others as quartz [384, 386, 388, 389]. Florinskaya and Pechenkina [386], who investigated the infrared spectra of quartz glass, reported that its structure may contain the Si–O–H group. It would be very significant to determine whether or not such groups are present in crystalline forms of silica, especially in quartz.

In connection with the study of the structure of quartz glass and its use in technology, a detailed investigation was also made of its strength characteristics and internal friction [390–393]. As a result of this it was established that the Young's and shear moduli of quartz glass have the minimal values at about 65°K. At a higher temperature (up to 1000°), the Young's modulus of this material increases approximately linearly with temperature. The strength of quartz glass increases with a rise in temperature and this was investigated up to 1200° [711].

Considering that quartz glass consists of atomic oscillators, Smyth and his co-workers [394] calculated the heat capacity of this material for different temperatures. With the exception of those for very low and very high temperatures, the results of the calculations agreed well with experimental data. Smyth thus showed that the frequency of the transverse oscillations of the oxygen atoms here were considerably less than for longitudinal ones (along the line of the Si–O–Si bonds). At moderate temperatures, the heat capacity of quartz glass is largely determined by the energy of transverse oscillations of oxygen atoms.

The thermal expansion of quartz glass depends on its preparation and shows hysteresis [7, 387]. The average values of the thermal expansion coefficient of quartz glass in relation to temperature, according to Sosman [7], are presented in Fig. 52. This figure shows that like crystalline modifications of silica, quartz glass has a negative thermal expansion over a certain temperature range. Quartz glass has a minimal volume at about 30°K [354].

The very low thermal expansion in comparison with low-temperature crystalline modifications of silica was previously explained

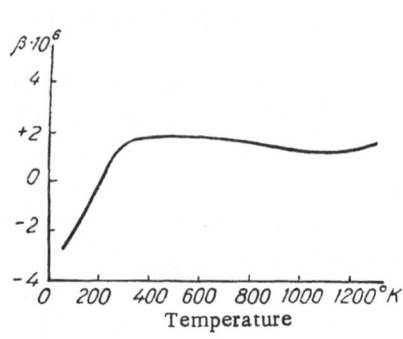

Fig. 52. Actual volume coefficient of thermal expansion of quartz glass in relation to temperature, according to Sosman (1927).

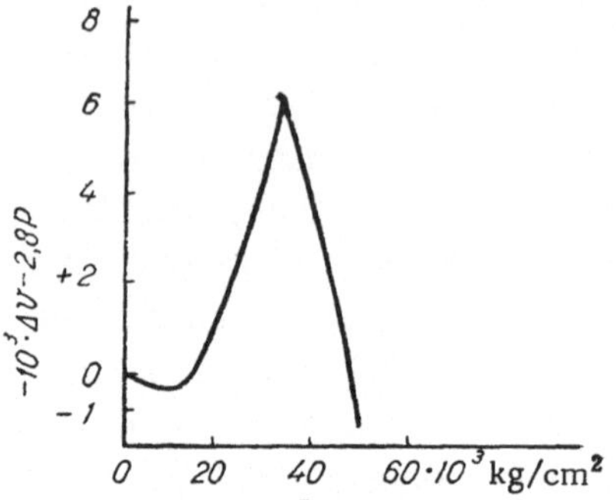

Fig. 53. Change in the volume of quartz glass in relation to pressure, according to Bridgman (1939).

by the impossibility of cooperative movements in the arbitrary loops of quartz glass. Since the Si—O bond is strong, changes in its length cannot be great. The high thermal expansion of crystalline forms of silica may occur only due to changes in the value of the bond angle. Thus, a 1° change in the angle of the Si—O bond in β-quartz from 150° leads to the same relative change in volume as after heating to 210° [395]. There is no definite structure in glass and therefore systematic series of rotations (or vibratory movements) are impossible.

The thermal expansion of quartz glass was investigated by Smyth [396] from the point of view of ideas on vibratory movements of atoms. By calculating the value of the free energy from the summed energies of the oscillators, Smyth found that the sign of the thermal expansion depends on whether there is a decrease or an increase in the frequency of the oscillator making the main energy contribution, i.e., transverse vibrations of oxygen atoms, with an increase in the interatomic distances. The packing of atoms in silica glass and high temperature crystalline modifications of silica is not the densest. In such structures there is an increase in the interatomic distances with an increase in the frequency of the oscillations of oxygen atoms (or their equivalents) and therefore the thermal expansion has a negative value. On reaching a definite value, the interatomic distances begin to decrease with an increase in the frequency of vibration of the oxygen atoms and the thermal expansion coefficient becomes positive. Processes analogous to this also occur in the very loose structures of the high-temperature modifications of silica and also in some silicates whose structures are similar to them. The packing of the atoms is denser in the low-temperature crystalline modifications of silica and the freedom for transverse oscillations of the oxygen atom is less here. As stated above, the thermal expansion of these forms of silica is determined by the change in bond angle.

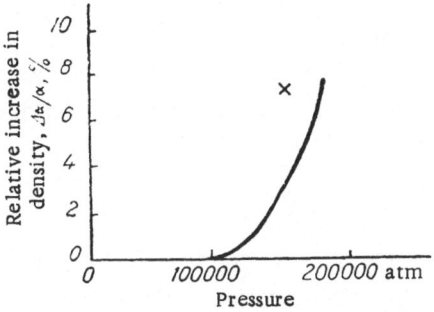

Fig. 54. Compression of quartz glass in relation to pressure at room temperature, according to Bridgman and Simon (1953). The cross marks the density reached at a temperature of 150°.

With an increase in pressure up to 31–33 kbars, there is a change of the second order, which results in a deviation from proportionality for the decrease in specific volume in relation to pressure (Fig. 53). Sosman [354] proposed the name of "infrapiezo glass" for normal quartz glass (lechatelierite) up to this change and the name "suprapiezo glass" (shortened to S-P glass) after the change at a pressure of 31–33 kbars. Smyth and his co-workers [398] showed that bending of the Si–O–Si bonds occurs when quartz glass is compressed. Some of these bonds are already bent even at atmospheric pressure. An x-ray study of quartz glass at high pressure showed that the compression process occurred without any sharply expressed changes in the structure. Bridgman and Simon [399] found that with an increase in pressure above 100,000 atm quartz glass began to contract strongly (Fig. 54), especially if the temperature was raised. The specific gravity of quartz glass reached 2.61 at 200,000 atm (the specific gravity of the original glass was 2.22). This is close to the specific gravity of quartz and considerably greater than that of tridymite and cristobalite, although the material obtained was completely amorphous. Sosman [354] proposed the name of "condensed" glass for such dense quartz glass. The halo on an x-ray pattern, obtained for such a quartz glass, gave an atomic separation of 4.11 A, which corresponds to the (111) plane of cubic cristobalite. This indicates no shortening of the Si—O bond, but only a change in the valence angle in the (SiO$_4$) tetrahedra.

Quartz glass compressed under a pressure of more than 100,000 atm does not recover its original volume elastically. Therefore, it is also possible to preserve thin disks of super-dense (condensed) quartz glass after removal of the pressure.

Thus, depending on the pressure, quartz glass may exist in three states.

The properties of quartz and quartz glass change [400] in opposite directions during irradiation with neutrons in a nuclear reactor (Table 19).

Table 19

Change in the Properties of Quartz and Quartz Glass Under the Action of Radiation from a Nuclear Reactor [400]

Radiation units	Quartz					Quartz glass		
	Specific gravity, g/cm³	Refractive index, D		$Ne-No$	Rotatory power, D, deg/mm	Specific gravity, g/cm³	Refractive index, D
		Ne	No				
0	2.6501	1.5534	1.5444	0.0090	21.5	2.2050	1.4589
1.3	2.6483	1.5530	1.5441	0.0089	—	2.2094	1.4597
2.9	—	1.5530	1.5440	0.0090	—	—	1.4611
49	2.5257	1.5272	1.5224	0.0048	14.5	—	—
59	—	—	—	—	—	2.2677	1.4673
1100 at a temperature of 250–300°	2.251	1.4666	1.4666	0.0000	0	2.251	1.4667

Note: The radiation dose unit $\approx 10^{-18}$ of the integrated beam of neutrons with an energy distribution of $\dfrac{K}{E}$, where K is a constant and $E = (0.01-1)$ Mev.

The specific refraction energy remains equal to 0.207 during this. The ruptures in the quartz lattice (of the metamictic disruption type) are healed by heating, though the properties are not completely restored. The refractive index of quartz glass begins to be restored with heating above 300°. When the materials were heated to 900° after irradiation in a nuclear reactor, the refractive index

(D), Ne, of quartz rose only to approximately 1.537 and that of quartz glass fell to 1.4583. The absorption spectra of both materials also changed during irradiation in a reactor, indicating deep conversions in them. The thermal effect of the conversion at 573° was removed completely from the thermograms of quartz. The Si–O distances in irradiated quartz and glass, like the O–O distances, are the same, while the Si–Si distance is less in the first case. The angles of the Si–O–Si bonds are 138 and 142°, respectively [677].

The properties of tridymite and cristobalite change during interaction with neutrons analogously to quartz and their specific gravity approaches 2.26 [354, 707]. X-ray investigation showed that the products obtained were amorphous.

Amorphous silica (glass) is obtained from quartz by the action of water at high pressure. According to x-ray analysis data, silica gel contains particles similar to cristobalite [380]. The SiO_2–H_2O system was investigated by Tuttle and England [318], who established that the melting point of silica is reduced in the presence of water or water vapor as stated above. The hypothetical phase diagram that they obtained is examined below.

In connection with the very great complexity of the conversions of different forms of silica and their exceptional importance in technology and geology, attempts were made to use a "model" (BeF_2) and analogs ($AlPO_4$, and $GaPO_4$) and also derivatives of the structures of different forms of silica. It was stated above that there is no analogy between SiO_2 and GeO_2. Investigation of the SiO_2 "model" BeF_2 [403] showed that the latter substance exists in the form of two modifications crystallizing like quartz (low- and high-temperature forms with a transition point at 220°) and high-temperature cristobalite which slowly changes into the first modification at 340–430°. The BeF_2 glass has a structure similar to that of SiO_2 glass [404]. No tridymite structure was observed for this compound.

Aluminum phosphate, $AlPO_4$, is also a "model" of silica [405, 406]. It was established that this substance has the following phase conversions:

Berlinite	$\xrightarrow{815°}$	Tridymite	$\xrightarrow{1025°}$	Cristobalite	$\xrightarrow{\geqslant 2000°}$	Melt
(a type of quartz)		form		form		
586		93° 130°		210°		
$\alpha \longleftrightarrow \beta$		$\alpha \longleftrightarrow \beta_2 \longleftrightarrow \beta_1$		$\alpha \longleftrightarrow \beta$		

The phase changes from one type of $AlPO_4$ to another proceed extremely slowly, but very rapidly within the limits of one type, as with silica. However, the analogy of $AlPO_4$ and SiO_2 is destroyed by differences of the structure of the melts [407]. Nonetheless, the existence of a "pure" tridymite type seems to have been established on the example of $AlPO_4$. Unfortunately, this modification of $AlPO_4$ has not been studied sufficiently as yet and therefore definite conclusions on it cannot be drawn, the more so as a tridymitelike modification was not detected for the next "model," SiO_2–$GaPO_4$ [408].

Among the silicates there are derivatives of all three main types of silica structure [409]: a) quartz type – α–eucryptite (α–$LiAlSiO_4$), $LiAlSi_2O_6$, etc.; b) tridymite type – nepheline ($NaAlSiO_4$), calcilite ($KAlSiO_4$), $K_2Mg\ Si_3O_8$, $K_2Fe\ Si_3O_8$, $K_2Mg_5\ Si_{12}O_{30}$, etc.; c) cristobalite type – carnegieite ($NaAlSiO_4$), Na_2CaSiO_4, etc.

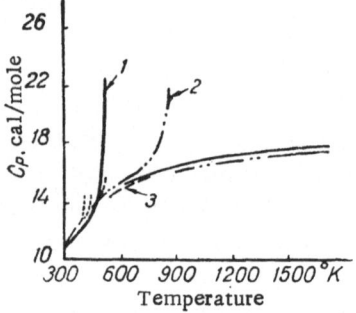

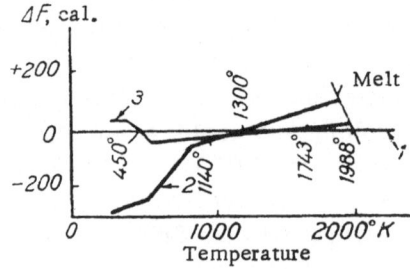

Fig. 55. Heat capacity of different modifications of SiO_2 in relation to temperature, according to Mosesman and Pitzer (1941). 1) cristobalite, 2) quartz, 3) tridymite.

Fig. 56. Free energy of formation of different modifications of SiO_2 from cristobalite in relation to temperature, according to Mosesman and Pitzer (1941). 1) cristobalite, 2) quartz, 3) tridymite.

It was stated above that β-spodumene, β-$LiAlSiO_4$, and μ-cordierite, μ-$Mg_2Al_4Si_5O_{18}$, are of the keatite type. It is possible that some chalcedonies are formed from the fibrous form of silicon, silica-W. No analogs of coesite are known as yet.

The existence of structural analogs of the different modifications of SiO_2 indicates that these structures are widespread. However, the polymorphism of structural analogs of silica, for example, $LiAlSiO_4$ and $KAlSiO_4$, is only just being studied.

Thus, these circumstances eliminate the possibility of drawing an unequivocal analogy between the polymorphism of silica and its structural "models" and the latter may be used only partially for indirect indications on this problem.

As is shown by Fig. 55, the polymorphic conversions of silica are accompanied by considerable changes in heat capacity, whose values were found more accurately by Mosesman and Pitzer [352] by determining the heat content of very pure preparations of quartz, tridymite, and cristobalite. The free energy of formation of the different normal modifications of silica from cristobalite, according

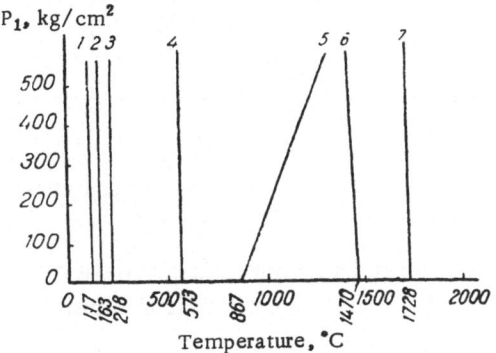

Fig. 57. Effect of pressure, P_1, on the temperature of modification changes of silica: 1) $\gamma \to \beta$-tridymite, 2) $\beta \to \alpha$-tridymite, 3) β-cristobalite $\to \alpha$-cristobalite, 4) β-quartz$\to\alpha$-quartz, 5) α-quartz $\to \alpha$-tridymite, 6) α-tridymite$\to\alpha$-cristobalite, 7) α-cristobalite$\to$melt.

to Mosesman and Pitzer, are presented in Fig. 56. This figure shows that the temperature for the conversion of α-quartz into α-cristobalite ($\Delta F = 0$) equals 1027° (1300°K). A quartz–cristobalite–melt ternary point exists at 1730° and a pressure of 1160 $\pm$ $\pm$ 500 atm (we corrected the temperature according to the latest determinations of the melting point of cristobalite).

Considering the data in Table 18, it is possible to represent the phase changes of the main modifications of silica (Fig. 57) in a system of $T—P$ coordinates with a scale and not arbitrarily, as has been done up to now. The vapor pressure of silica was found to be so low that it cannot be shown in the diagram. Figure 57 shows that only the slope of the line for the conversion of α-quartz into α-tridymite is very considerable. The others differ little from the vertical (see data in Table 18).

Taking the classical data on the phase changes of quartz, tridymite, and cristobalite, since the objections to them still require

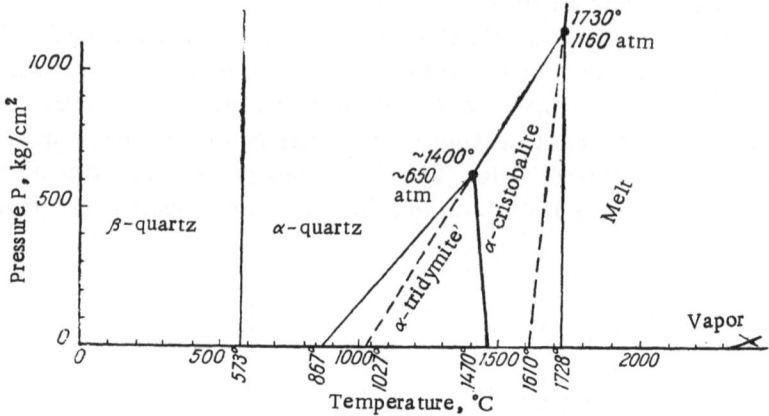

Fig. 58. Phase diagram of SiO₂ according to Mosesman and Pitzer with additions according to the latest data.

further investigations, and considering the data of Mosesman [352], Keith and Tuttle [309], Kamentsev [150], Sosman [354], and others,

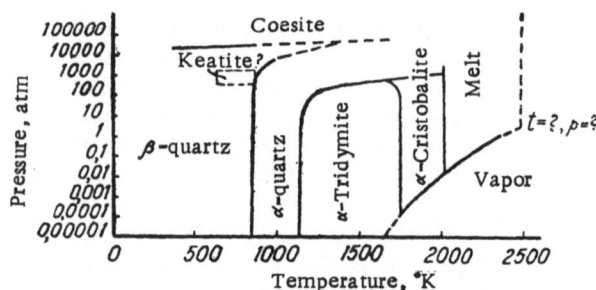

Fig. 59. Phase diagram of SiO₂ (in the presence of a mineralizer–melt).

it is possible to build up a thermodynamically based phase diagram for silica to replace the obsolete diagram of Fenner. Such a diagram is shown in the normal coordinates in Fig. 58. The limits of the conversion of quartz into cristobalite and of quartz fusion are indicated by a dotted line. According to classical ideas, these are metastable changes.

However, it is difficult to indicate the vapor region and also even the approximate positions of the keatite and coesite regions

in normal coordinates. This is eliminated by the use of semiloga-
rithmic coordinates. Using these, we constructed the more complete
phase diagram of silica shown in Fig. 59. The critical points for
the melt have not been established as yet. The position of silica-W

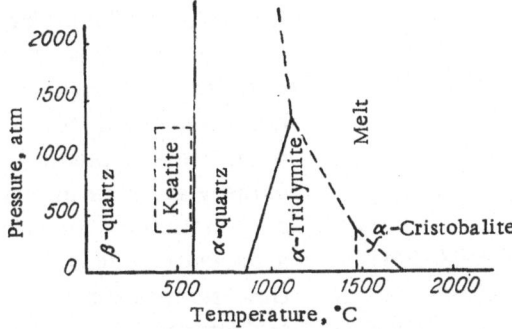

Fig. 60. Phase diagram of SiO_2 in the presence of
water, according to Tuttle and England (1955).

cannot yet be shown on Figs. 58 and 59 due to the absence of appropriate data. For simplicity, the phase conversions of tridymite
and cristobalite are not shown on Figs. 58 and 59.

Scheme of SiO_2 conversions

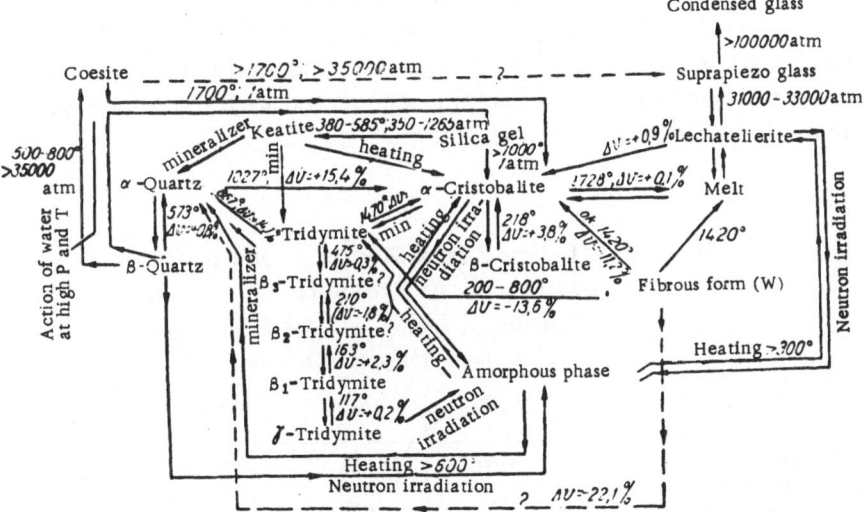

The melting points of silica modifications are lowered in the
presence of water vapor. According to a preliminary phase diagram
of silica in the presence of water, put forward by Tuttle and England

[318] and presented in Fig. 60, cristobalite is not formed at a pressure of more than approximately 400 atm and tridymite fuses into a hydromelt. If the pressure is more than 1400 atm, then tridymite is not formed either. Only quartz remains under these conditions and it melts at 1000–1100° into a hydromelt. The region at which keatite exists is shown provisionally on the diagram in Fig. 60 as complete data for it have not yet been determined. In exactly the same way, it is impossible to arrive at conclusions on whether there is a pressure region in which β-quartz (low-temperature) can fuse into a hydromelt.

Considering all the data presented on the properties of different forms of silica, we attempted to present a scheme of SiO_2 conversions (see p. 135), which is very complex.

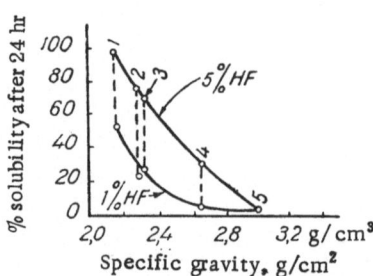

Fig. 61. Solubility of different SiO_2 modifications in HF, according to Coes (1953): 1) SiO_2 glass, 2) tridymite, 3) cristobalite, 4) quartz, 5) coesite.

The scheme presented is not yet accurate in many sections since there are no complete experimental data. Nonetheless, it shows the reason for the great variety of processes in which silica participates in nature and technology. Consequently, it is still necessary to examine some problems connected with the solubility of silica, especially in water and steam.

Table 20

Solubility of Different Silica Modifications in Water
in Relation to Pressure

Silica modification	Solubility at 400°, g per kg of water, at the following pressures, bars				
	450	500	1000	1500	2000
Quartz	—	1.10	1.65	2.00	2.20
Cristobalite ..	1.21	—	1.69	2.06	—
Tridymite ...	1.34	—	1.96	2.62	—

In reactions of the different forms of silica, it is usually found that quartz is least and tridymite most reactive. This was established by Budnikov and Krech [410] by a study of the interaction of SiO_2 modifications with chlorine at high temperatures and by Thümmler and Klügel [411] in reactions with $CaSO_4$.

Coes [369] established that the solubility of different forms of silica in aqueous hydrofluoric acid decreased approximately with an increase in their density (Fig. 61). Quartz glass dissolves especially readily and coesite is almost insoluble in HF.

Wyart and Sabatier [412] determined the solubility of quartz, tridymite, and cristobalite in water at 400° and various pressures. The solution time was up to 10 days. (Table 20.)

As can be seen from Table 20 and Fig. 61, the results of experiments by Wyart and Sabatier and Coes are identical: quartz is the least and tridymite the most soluble of all the normal crystalline modifications of silica.

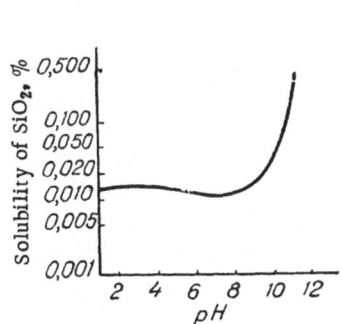

Fig. 62. Solubility of silica in water in relation to pH, according to Alexander, Heston, and Iler (1954).

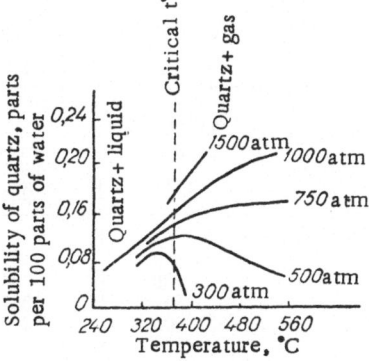

Fig. 63. Solubility of quartz in water and steam in relation to temperature and pressure, according to Walker (1953).

Alexander and his co-workers [413] studied the water solubility of amorphous silica. The solubility of amorphous silica is 0.010–0.012% at 25°. The temperature dependence of the water solubility of amorphous silica up to 200° is practically linear and may be expressed approximately by the equation

$$C = 0.382 \ (13.6 + t) \cdot 10^{-3},$$

where C is the concentration of silica in solution, %, and t is the temperature, °C.

Silica is presented in solution both in the form of H_2SiO_3 and as H_4SiO_4. The H_2SiO_3 concentration depends on the pH of the solution (Fig. 62), while the H_4SiO_4 concentration is practically independent of it.

In connection with the solubility of silica in liquids liberated by the tissues of an organism during silicosis, new investigations were made of the solubility of this substance in water and physiological liquids [414]. It was established that when silica powder is steeped in water, the concentration of silicic acid first increases sharply and then slowly. The rapid solution of silica occurs at the expense of corners, sharp edges, and amorphous layers, which were found to be 200–1500 A; their presence in quartz was proved by an electron diffraction method [415].

Table 21

Solubility of Quartz and Quartz Glass in Water Vapor
at High Pressure

Temperature, °C	Water vapor pressure, kg/cm²	Solubility, g per million g of water		Partial pressure of SiO₂ in vapor during solution of quartz, atm
		quartz glass	quartz	
400	35.2	—	1.04	—
	70.3	—	3.1	0.00021
	140.6	34	5.2	0.00071
	351.6	1381	637	0.217
	703.1	2429	1259	0.856
	1054.7	—	1501	1.580
500	35.2	—	3.77	—
	70.3	—	14.2	0.00096
	140.6	—	35.7	0.0033
	351.6	346	216	0.073
	703.1	2539	1389	0.92
	1054.7	4179	2472	2.65

In developing a hydrothermal method for growing large quartz crystals, Walker [321] investigated the solubility of quartz in water and water vapor, in relation to temperature and pressure. Figure 63 shows that the solubility of quartz increases with pressure. A maximum close to the critical temperature of water is observed on the curves of solubility against temperature at pressures up to 500 atm. Mosebach [416] attempted to consider the hydrothermal solubility of quartz as a heterogeneous gas equilibrium.

The solubility of quartz and quartz glass in water vapor at high pressure was investigated in detail by Morey et al. [417, 418, 485, 489]; some of their results are presented in Table 21.

The table shows that the solubility of quartz in water vapor increases with an increase in pressure. By a graphical representation of the data in Table 21, it is possible to prove that the solubility of quartz is greater at 400° than at 500° at pressures of 141–633 kg/cm^2 and the reverse at other pressures. In relating the solubility of silica to vapor density, the solubility increases with temperature at all pressures. The greater solubility of quartz glass is connected with the value of its free energy. Silica in water vapor is probably present as H_4SiO_4 or another hydrated form.

According to Gruner [419], the solubility of chalcedony in water at 300° is about 0.11% and that of quartz is 0.06–0.13%.

The solubility of silica in superheated steam is of great importance in geology in connection with pneumatolysis phenomena and also in steam power engineering (steam boilers and high-pressure turbines). Substances dissolved in the steam must be deposited on the blades of steam turbines and these include: quartz at 380–150°, cristobalite at 150–80°, and amorphous silicic acid at 150–40° [419].

The adsorption properties of quartz were investigated by Sarakhov [679] and Störber [680]. Quartz catalytically decomposes isopropyl alcohol into propylene and water [680].

Other properties of various forms of silica and their practical use have been described in our own literature [6, 10–12, 15, 16, 18, 20, 387, 702, and others] and it is therefore unnecessary to dwell on them. We will only note that the range of uses of SiO_2 in different forms is very great in diversity and number. Thus, for example, optical apparatuses are made from quartz and quartz glass and the elastic properties of quartz are used for the construction of piezoelectric apparatuses. Silica is also a basis for glass manufacture and for many constructional materials and Dinas refractories, which are made in many hundreds of thousands of tons per year. As it is the most important rock-forming mineral, silica is also of very great significance in geology. It can be stated truly that the further study of this substance, which is of exceptional interest in all respects, will be of even greater benefit to mankind.

Si–S SYSTEM. Silicon disulfide, SiS_2, was prepared by Berzelius. Sabatier [428] determined its heat of formation. The properties of this compound were studied by a series of investigators [420–424]. The method of Tiede and Thimann [422] is usually used for the preparation of silicon disulfide and this consists of the reaction of SiO_2 with Al_2S_3 at 1200–1300° in a stream of nitrogen, since the crystallization of this substance proceeds poorly in vacuum. The silicon disulfide formed evaporates and precipitates in cold parts of the reaction zone in the form of needlelike crystals of hexagonal cross-section with high birefringence and straight elongation. The reaction does not proceed quantitatively due to the formation of silicon monosulfide, SiS [425, 426]. Kohlmeyer and Retzlaff [426] established that silicon disulfide may be obtained by the reaction

$$Si + 2S = SiS_2$$

at temperatures of 1100–1200° in an alumina or carbon crucible. Silicon disulfide is the primary reaction product of sulfur and silicon. Silicon monosulfide is only formed from the disulfide [427].

Silicon disulfide is obtained either in a white or a yellow form and this is apparently connected with the purity of the preparation. Its heat of formation was determined by a series of investigators, who obtained very varied results, as can be seen from the following:

Author	$-\Delta^H$form SiS_2, kcal/mole
Sabatier [428]	
for white preparation	38.8
for yellow preparation	54.0
Kubaschewski and Evans [429]	39.0±8.0
Gabriel and Alvarez-Tostado [425]	59.9±0.5
Rocquet and Ancey-Moret [430]	43.2

The last value is most accurate.

The structure of silicon disulfide was investigated by Zintl and Loosen [431] and also by Büssem, Fischer, and Gruner [432], who established that this compound belongs to the rhombic syngony (see Table 2) and has the structure illustrated in Fig. 64. The structure is characterized by the presence of somewhat deformed (SiS_4) tetrahedra, connected by edges, as in fibrous silica-W, and forming infinite chains. In connection with this, silicon disulfide crystals also have a fibrous structure [431]. Its known properties are presented in Table 2. The melting point of this compound was determined by Tiede and Thimann [422].

Silicon disulfide is decomposed by water, especially water vapor, with the formation of SiO_2 and H_2S. On hydrolysis it forms meta-

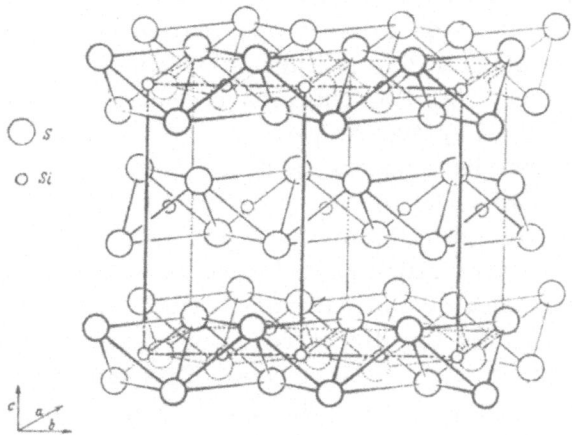

Fig. 64. Crystal lattice of SiS_2; space group D_{2h}^{20}—Icma
$a = 5.60$ A, $b = 5.53$ A, $c = 9.55$ A, $z = 4$; a doubled
elementary cell (in the direction of the b axis) is shown.

silicic acid [433] and the hydrolysis is accelerated by the presence of alkalies. Silicon disulfide is oxidized by atmospheric oxygen only on heating.

Tiede and Thimann [422] obtained phosphorescent silicon disulfide by activating it with carbon.

Silicon monosulfide, SiS, has been investigated less than the disulfide. It appears as a white amorphous mass or yellowish needles. The existence of this compound was established by absorption spectra in the ultraviolet region [434, 435]. It was thus found [435] that the Si–S distance is about 1.93 A, i.e., about 11.5% less than for a purely covalent bond. Silicon monosulfide is obtained from the disulfide and silicon at high temperatures in the gas phase and also from FeS and silicon at 980° [430]. The boiling point of silicon monosulfide at

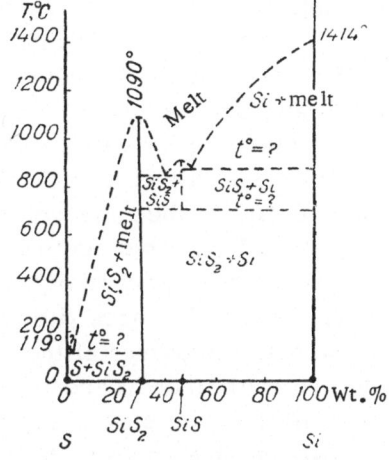

Fig. 65. Hypothetical phase diagram of the Si–S system.

atmospheric pressure is 940°. With slow cooling, it decomposes to silicon disulfide and silicon [427].

Figure 65 shows a hypothetical phase diagram of the Si–S system, which summarizes present information on silicon sulfides. The structure of this phase diagram is analogous to that of the Si–O system (see Fig. 42). There have been no systematic investigations of the S–Si system as yet.

Silicon sulfides have no practical use as yet.

Si–Se SYSTEM. The Si–Se system has been studied even less than the Si–S system. Sabatier [436] obtained the compound $SiSe_2$ by passing H_2Se over silica heated to redness, but his preparation was not pure. Gabriel and Alvarez-Tostado [425] obtained pure silicon diselenide by the direct reaction of silicon and selenium at 650° in a sealed tube of quartz glass. This substance was white. Its heat of formation was 1.33 ± 0.5 kcal/mole.

Silicon diselenide monocrystals up to two cm in length were first obtained by Weiss [437] by heating silicon and selenium in vacuum for a week. They were freed from selenium in high vacuum at 900° and then this selenide was sublimed at 1100°. Silicon diselenide appeared as clear, colorless, flexible fibers with a low hardness. At the same time, a small amount of the monoselenide, SiSe, was obtained as a yellowish brown powder. Silicon diselenide belongs to the rhombic system (see Table 2). Its structure is the same as SiS_2 [437]. The Si–Se distance in silicon diselenide equals 2.23 A and the Se–Se distances in each $SiSe_4$ tetrahedron are 3.39, 3.73, and 3.77 A. Thus, this tetrahedron is irregular. The closest Se–Se distance for neighboring chains equals 3.87 A.

If crystalline silicon diselenide is heated in an evacuated quartz tube to 1060°, then the compound melts under the pressure of its vapors and on rapid cooling, it gives a clear, light yellowish glass with a specific gravity of 2.95 [438]. An x-ray investigation of this glass gave only three diffuse halos (6.4; 3.1; 1.7 A). The change of the glass into crystalline silicon diselenide must be accompanied by a very great decrease in volume, namely 23.7%. The glass readily crystallizes with slow cooling.

Silicon diselenide is decomposed at room temperature by atmospheric oxygen with the formation of SiO_2 and selenium [425]. Water decomposes it with the formation of SiO_2 and H_2Se. Its decomposition is almost complete after a short period, even in moist air, when SiO_2, H_2Se and red selenium are formed [437]. Silicon diselenide purified by sublimation fluoresces with a yellowish green light in x-rays.

Due to its smaller reactive surface, the glassy form of silicon diselenide is more stable chemically than the crystalline form. However, water decomposes it at room temperature and it dissolves in concentrated sulfuric acid with prolonged heating to give a green color. An aqueous solution of NaOH rapidly decomposes glassy silicon diselenide with the formation of a red solution of sodium polyselenide.

The existence of silicon monoselenide was demonstrated at 800–1000° by ultraviolet absorption [434]. This compound has not been investigated yet.

The phase diagram of the Si–Se system has not been worked out yet. Its structure is apparently similar to that of the Si–S system.

Silicon selenides have no practical use.

Si–Te SYSTEM. Silicon ditelluride, $SiTe_2$, was obtained by synthesis from silicon and tellurium at 1050–1070° in high vacuum [439] and appeared as red needlelike and flaky crystals. The existence of considerable anisotropy in the hardness of these crystals was established [440]. Silicon ditelluride sublimes without melting and with partial decomposition when heated to 1220°. Zones of silicon ditelluride, tellurium, and silicon monotelluride are formed.

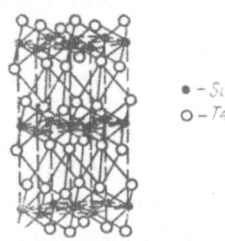

Fig. 66. Crystal lattice of the hexagonal modification of $SiTe_2$.

Silicon ditelluride crystallizes in the hexagonal system like the hexagonal modification of CdI_2 (Fig. 66). The lattice dimensions are presented in Table 2. The Si–Te distance in this modification of silicon ditelluride equals 3.04 A, which is greater than the sum of the ionic radii (2.50 A) and corresponds to the sum of the atomic radii. This indicates the presence of a metallic bond in this modification [440].

In addition to the hexagonal modification described, the preparation of silicon ditelluride accidentally yielded a new form of it, namely, a fibrous one, which is probably isomorphous with SiS_2 and $SiSe_2$ [439], but which has not been studied yet.

Red crystals of silicon ditelluride rapidly decompose in air with the formation of H_2Te. Dilute nitric acid slowly decomposes it at room temperature and rapidly on heating. Concentrated sulfuric acid reacts with silicon ditelluride, but the reaction stops due to the formation of a protective film. Concentrated solutions of alkali rapidly decompose it with the formation of a red solution [440].

Silicon monotelluride, SiTe, belongs to the cubic system [439]. The existence of this compound was also proved by the ultraviolet absorption spectrum at 800–1000° [434].

The phase diagram of the Si–Te system has not been worked out at all.

Silicon tellurides are not used practically as yet.

Si–Po SYSTEM. This system has not been studied at all. In analogy with sulfur, selenium, and tellurium, it can be assumed that the compounds $SiPo_2$ (6.2% Si) and SiPo (11.8% Si) exist. If these compounds exist, it is to be expected that their melting points will not be higher than those of the analogous compounds of silicon with selenium and tellurium (i.e., not above 1100°).

Systems Formed by Silicon with Transition of Metals of Group IV

The Ti–Si, Zr–Si, and Hf–Si systems have not been studied to the same extent. While the first two are known in detail, the phase diagram of the latter has not even been worked out in a preliminary form. Nonetheless, it is possible to trace (see Table 2) a very great analogy in the properties of titanium, zirconium, and hafnium silicides, which is caused by the identical structure of their external s electron shells ($2s$ electron) and the similar ratios of their atomic radii to the radius of the silicon atom (1.47, 1.60, and 1.59, respectively, for titanium, zirconium, and hafnium). All the silicides of transition elements of Group IV are typical intermetallic compounds. They are characterized by a metallic luster, electroconductivity, and quite a high hardness (see Table 2).

The structures of the phase diagrams of the Ti–Si and Zr–Si systems show a certain, though not complete analogy, which includes a maximal congruent melting point of the compounds Me_5Si_3 and incongruent melting of the monosilicides MeSi. The silicides Me_5Si_3 and MeSi melt above and the silicides $MeSi_2$ below the melting point of the metal. The structure of Me_5Si_3 (type $D8_8 - Mn_5Si_3$) is characterized by the presence of four double electron layers in the lattice [604, 605].

In the phase diagrams of Si–Ti, Si–Zr, and Si–Hf, the melting points of the pure metals are corrected according to the latest

determinations of Deardorff and Hayes [441], which were: titanium –
$1668 \pm 10°$, zirconium – $1855 \pm 15°$, and hafnium – $2220 \pm 30°$. The
first two temperatures are close to those reported previously,
whereas the latter is almost 250° higher, which is probably explained by a difference in the purity of the hafnium and the accuracy
of the procedure for determining its melting point. The preparation
of silicides of transition elements of Groups IV, V, and VI were
described above and therefore is not presented in this section.

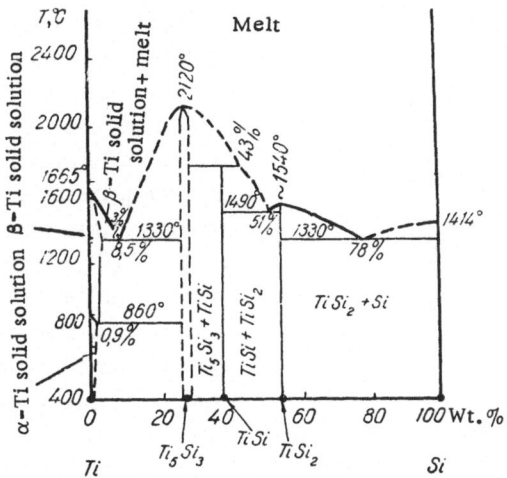

Fig. 67. Phase diagram of the Ti–Si system according to Hansen, Kessler, and
McPherson (1952).

The heats of formation of some silicides of transition metals
of Groups IV, V, and VI (Ti_5Si_3, TiSi, $TiSi_2$, Zr_5Si_3, ZrSi, $ZrSi_2$,
$ThSi_2$, VSi_2, Ta_5Si_3, $TaSi_2$, $MoSi_2$, WSi_2) were recently determined
by Robins and Jenkins [624]. Other properties of silicides of Group
IV–VI transition metals were compared by Kieffer and Benesowsky
[664].

Ti–Si SYSTEM. Titanium silicide was first obtained by Moissan
(1895) but he did not investigate its composition. A silicide with the
composition Ti_2Si_2 (22.6 wt.% Si) was obtained by Levy (1895) by
the action of $TiCl_4$ on silicon at red heat. Titanium disilicide, $TiSi_2$,
was first obtained in a pure form by aluminothermy from quartz,
potassium fluotitanite, sulfur, and aluminum by Hönigschmid [442].
Askenazy and Ponnaz [443] described the compound Ti_2Si_3. Then
many investigators studied titanium silicides [444, 445].

The phase diagram of the Ti–Si system was worked out by Hansen, Kessler, and McPherson [446]. The solubility of silicon in titanium was also determined previously by Graigheat and his co-workers [447], but the values they obtained were found to be low and were corrected by Hansen.

Titanium (99.7% purity) and silicon (99.9% purity) were used for working out the phase diagram of the Ti–Si system. The investigators used thermal analysis (up to 1600°) and also x-ray and metallographic investigations of the alloys.

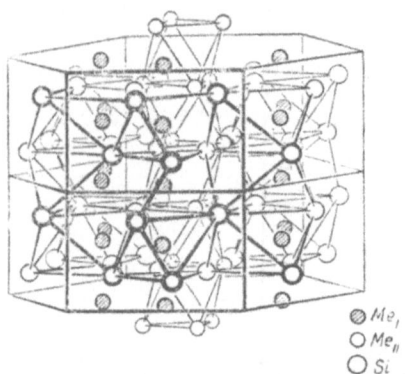

The results obtained are presented in Fig. 67. The figure shows that the structure of the phase diagram of the Ti–Si system and no special explanations are required. There are three intermediate phases in this system: Ti_5Si_3 (region of homogeneity at 26 – 28% Si), TiSi, and $TiSi_2$. The compound Ti_2Si was not obtained [448]. The maximal melting point (2120°) was not obtained at the stoichiometric silicon content of Ti_5Si_3 (26.01%), but at 27.79% Si. According to Cerwenka, the melting point of $TiSi_2$ is not 1540, but 1460° [26].

Fig. 68. Crystal lattice of Me_5Si_3 of the $D8_8$ type (hexagonal modification); space group D_{6h}^3 — $C6/mcm$. The following silicides crystallize in this way: Ti_5Si_3, Zr_5Si_3, V_5Si_3, Nb_5Si_3, Ta_5Si_3, Cr_5Si_3, Mo_5Si_3, W_5Si_3, Mn_5Si_3 and Fe_5Si_3. The presence of carbon or nitrogen is necessary for the stabilization of this structure in the case where $Me = V$, Nb, Ta, Cr, Mo, and W.

The solubility of silicon in titanium is about 0.1% at room temperature. The presence of silicon in a solid solution in titanium reduces the temperature of the $\beta \rightleftarrows \alpha$-Ti phase transition.

Hansen et al. [446] established that alloys of titanium and silicon may be cold rolled up to an Si content of 2% with a linear increase in the hardness of the alloy in relation to the silicon content. The microhardness increased from 80 kg/mm² for pure titanium to 160 kg/mm² for an alloy with 2% Si. With quenching from a temperature of 1100°, the microhardness of this alloy equaled 260 kg/mm². A summary of the properties of titanium–silicon alloys was presented by Eremenko [449].

The crystal structure of Ti_5Si_3 was found [450] to be isomorphous with Mn_5Si_3 (Fig. 68). The dimensions of the elementary cell are presented in Table 2.

The structure of TiSi is rhombic (Table 2). The structure of $TiSi_2$ was investigated by Loves and Wallbaum [451], who found that here there are layers of a hexagonal lattice of silicon atoms, in whose centers are titanium atoms (Fig. 69). The interatomic distances in a layer are as follows: Si–Si – 2.75 A, Ti–Si – 2.75 A; the distances between atoms of nearest layers are: Si–Si – 2.54 A, Ti–Ti – 3.19 A, and Ti–Si – 2.54 A. The size of the elementary cell is given in Table 2. However, Cotter et al. [452] found that titanium disilicide is dimorphous. In the preparation of this compound by aluminothermy from TiO_2 and SiO_2 and also by sintering titanium hydride and silicon, a new modification is obtained, which is isomorphous with $ZrSi_2$ and $HfSi_2$

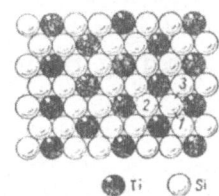

Fig. 69. Packing of atoms in the $TiSi_2$ lattice.

(see Table 2). The phase transition point has not been established yet, but it is apparently above 1200°.

None of the three titanium silicides change into the superconducting state down to 1.2°K [453]. Titanium disilicide is very friable at room temperature.

Titanium disilicide does not change in air. It is slowly oxidized when heated in air to red heat. Chlorine reacts with it under these conditions explosively, while bromine and iodine do so slowly [23]. The acids H_2SO_4, HCl, HNO_3, and $HClO_4$, 50% NaOH and KOH, and also molten (200–300°) $KHSO_4$ do not decompose titanium disilicide. However, it is decomposed by molten borax, NaOH, and KOH [452].

This compound dissolves in hydrofluoric acid (diluted 1:1). Titanium disilicide reacts with boron at 1450° with the formation of TiB_2 [27]. The silicide Ti_5Si_3 does not react with vanadium, niobium, tantalum, chromium, molybdenum, or tungsten at 1500°. This compound does not react either with molten copper, silver, or nickel. These metals may therefore act as binders in articles of Ti_5Si_3.

As yet, titanium silicides have no separate use. Alloys of titanium containing a small amount of silicon and also the silicide Ti_5Si_3 are of interest for the preparation of heat-resistant (refractory) articles.

Zr–Si SYSTEM. The first zirconium silicides were synthesized by Wedekind [454], who reduced ZrO_2 with silicon in an electric furnace. He obtained balls with a metallic luster. Fracture of these balls revealed silvery crystals and Wedekind proposed that their composition corresponded to the formula ZrSi.

Zirconium disilicide, $ZrSi_2$, was synthesized by Wedekind [455] by the direct interaction of zirconium and silicon in vacuum at 1000° and also by Hönigschmid [442] by reduction of K_2ZrF_6 in the presence of K_2SiF_6 at 1300° with aluminum.

A review of the first syntheses of zirconium silicides was presented by Hönigschmid [23] and Baroduc-Müller [444]. Zirconium

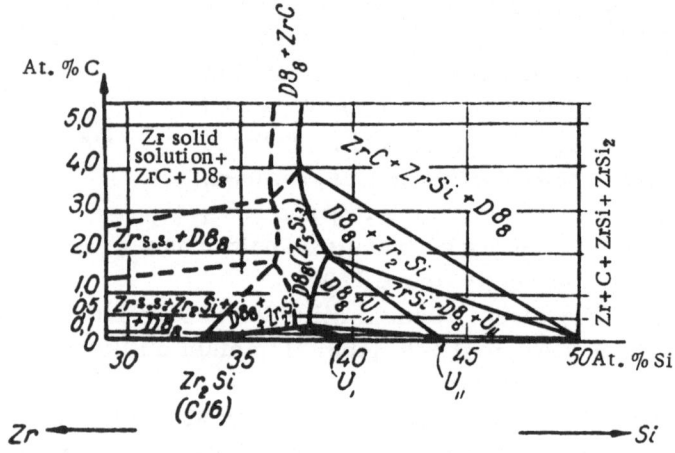

Fig. 70. Phase fields in the Zr–Si–C system according to Nowotny, Lux, and Kudielka (1956).

disilicide is formed as an intermediate product in the reduction of $ZrSiO_4$ with carbon [456]. This compound may be obtained as a layer on metallic zirconium from $SiCl_4$ vapor in the presence of hydrogen at 1100–1500° and also by hot-pressing of mixtures of zirconium and silicon.

The phase diagram of the Zr–Si system was worked out by Lundin, McPherson, and Hansen [457], who used 99.8% zirconium and 99.99% silicon for this. Thermoanalysis and microscopic investigations were used to find the intermediate phases: Zr_4Si, Zr_2Si, Zr_3Si_2, Zr_4Si_3, Zr_6Si_5, ZrSi, and $ZrSi_2$. Additional investigation by Kieffer, Benesowsky, and Machenschalk [458] with the aid of x-ray analysis showed that the Zr–Si system contained only the following binary phases: Zr_2Si, Zr_5Si_3, ZrSi, and $ZrSi_2$.

The results of an investigation on the effect of carbon content on the stabilization of different phases of the Zr–Si system were published recently [634]. It was shown (Fig. 70) that even at 0.2 at. % C, the two intermediate phases lying between Zr_5Si_3 and ZrSi (u_I and u_{II}) disappear and only Zr_5Si_3 is stabilized in the form of crystals of the $D8_8$ type, which are absent with zirconium iodide (very pure). The u_I phase corresponds to Zr_3Si_2, isotypic with U_3Si_2, and the u_{II} phase, which has the composition $Zr_{57.2}Si_{42.8}$ (18.72 wt.% Si), is approximately Zr_6Si_5 (20.42 wt.% Si).

The phase diagram of the Zr–Si system, corrected in accordance with these data, is presented in Fig. 71. The Zr_5Si_3 phase has

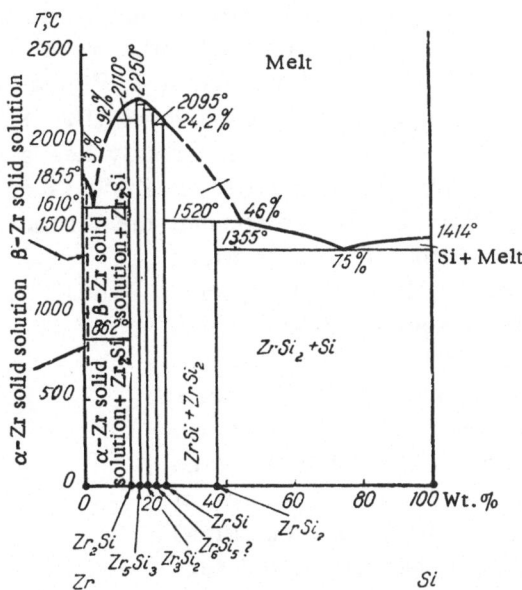

Fig. 71. Phase diagram of the Zr—Si system according to Lundin, McPherson, and Hansen (1953), with corrections according to Kieffer, Benesowsky, and Machenschalk (1954) and also Nowotny, Lux, and Kudielka (1956).

been left on it, since this part of the diagram has still not been studied sufficiently and we cannot exclude the possibility that the phases Zr_3Si_2 and even Zr_6Si_5 will be found to be only different forms of Zr_5Si_3, as was proved for the corresponding silicides of transition metals of Groups V and VI (see below).

The solubility of silicon in α-Zr is less than 0.1% at 860° and in β-Zr, less than 0.2% at 1610°. The effect of silicon content in solid solution on the $\alpha \rightleftarrows \beta$-Zr conversion temperature is apparently insignificant, though it has not been established accurately yet. The solubility of zirconium in silicon is considerably less than 5%, though this also has not been established accurately [457]. The limits of homogeneity of binary phases of the Zr–Si system have not been determined. The other details of the phase diagram of this system can be seen in Fig. 70. The main properties of zirconium silicides are presented in Table 2.

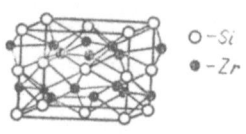

O – Si
● – Zr

Fig. 72. Crystal lattice of Zr_2Si (CuAl$_2$ type).

The crystal structure of Zr_2Si was investigated by Schachner, Nowotny, and Machenschalk [459], and Pietrakowsky [460], who obtained the same results apart from small differences in the dimensions of the elementary cell (the results of Schachner are given in Table 2; Pietrakowsky's values are as follows: $a = 6.6120$ A, $c = 5.2943$ A).

This phase apparently has a narrow region of homogeneity [460]. The structure of Zr_2Si is of the same type as CuAl$_2$ (Fig. 72). In this compound, the zirconium atoms are arranged in the form of a chain along the c axis. Each silicon atom is surrounded by eight zirconium atoms, which form a deformed Archimedian antiprism.

The structure of Zr_5Si_3 was determined by Schachner [459]. It was found to be the same as that of Ti_5Si_3 (see Fig. 68 and Table 2). The structure of Zr_3Si_2 is isotypic with U_3Si_2 (see below) while that of Zr_6Si_5 has not been determined yet.

A hexagonal cell with $a = 7.005$ A and $c = 12.772$ A was first proposed

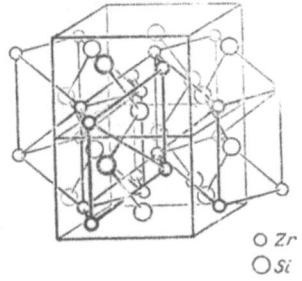

O Zr
O Si

Fig. 73. Crystal lattice of ZrSi (FeB type).

[457] for ZrSi, but a check showed [461] that the structure of this compound is rhombic of the FeB type (Fig. 73) and the size of its cell is given in Table 2; the silicon atoms in the crystal lattice form a zigzag chain.

Zirconium disilicide, $ZrSi_2$, has a gray color and crystallizes in the rhombic syngony, but not in the same group as $TiSi_2$ (see Table 2). The structure of zirconium disilicide was the subject of a series of investigations [461–464]. It was established that it consists of layers of zirconium and silicon atoms (Fig. 74), parallel to the (010) plane. The silicon atoms form chains parallel to the two crystallographic axes. Dimorphism has not been established for this compound [452].

In a compact form, zirconium disilicide is quite stable to oxidation during heating in air. Mineral acids, apart from hydrofluoric, do not act on it [23] and likewise 50% solutions of NaOH and KOH [452]. However, molten alkalies decompose it.

On heating, zirconium disilicide rapidly decomposes platinum in the presence of molten borax [452].

It ignites on being heated in oxygen. This silicide reacts vigorously with chlorine and other halides when heated to red heat and with fluorine, even with a very slight rise in temperature.

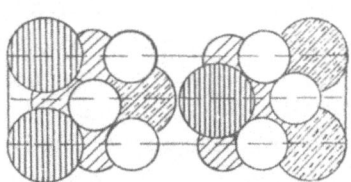

Fig. 74. Crystal lattice of $ZrSi_2$ according to Nardy-Szabo (1937). The large circles represent Zr and the small ones, Si. The uncrosshatched circles are in the plane of the drawing and the crosshatched ones, 1.835 A above and below.

Zirconium disilicide powder readily forms colloidal systems, which have a grayish brown color.

Zirconium disilicide reacts with boron at 1650° [27] to form ZrB_2. The technical properties of this compound were described also by Miller [465]. The properties of other zirconium silicides have been studied little as yet. Treatment of Zr_5Si_3 with ammonia at 1400–1800° forms ZrN [634]. The silicide ZrSi does not react at 1500° with vanadium tantalum, chromium, molybdenum, or tungsten or with molten copper, silver, or nickel [662]. Therefore these metals may be used as binders for zirconium monosilicide.

Zirconium silicides have no practical use as yet. Zr_5Si_3, which has the highest melting point (2250°), is of interest in refractory technology.

It should be noted that the ternary system Be–Zr–Si forms a binary hexagonal silicide ZrBeSi ($a = 3.71$ A, $c = 7.19$ A, $z = 2$, specific gravity 4.95).

Hf–Si SYSTEM. Post and his co-workers [466] synthesized two hafnium silicides, HfSi and $HfSi_2$, in vacuum at 1200° and in a helium atmosphere at 1500° and the properties of these compounds that have been studied are presented in Table 2. These hafnium silicides are isomorphous with zirconium silicides. Cotter [452] repeated the synthesis of $HfSi_2$ and confirmed the data of Post. Like titanium and zirconium disilicides, $HfSi_2$ is insoluble in mineral acids, with the exception of hydrofluoric acid, and in 50% NaOH and KOH, but is decomposed by molten alkalies, borax, and $KHSO_4$.

The Hf–Si system has not been investigated yet. In addition to the silicides mentioned, it is probable that Hf_2Si (7.3% Si) and Hf_5Si_3 (8.6% Si) also exist. Their melting points should apparently be of the order of 2400–2500 and 2500–2600°, respectively, and that of HfSi, of the order of 2400°. Determining the phase diagram of the Hf–Si system is of interest from the point of view of the technology of special refractories.

Hafnium silicides have no practical application as yet.

Systems Formed by Silicon with Transition Metals of Group V

The V–Si, Nb–Si, and Ta–Si systems have much in common, especially the latter two. As in systems with transition metals of Group IV, this is caused by the similarity in the structure of the outer electron shells, which is especially marked for niobium and tantalum, with the exception only of the presence of one electron ($O\,5s$) in the outer shell of niobium and two ($P6s$) in the case of tantalum. The previous levels are partially (vanadium) or completely (niobium and tantalum) unfilled. The atomic radii of vanadium, niobium, and tantalum (1.35, 1.47, and 1.47 A, respectively) are very similar [467]. In connection with the large values of these radii, all three systems form silicides of complex structure, especially at a silicon content of 40–50 at. %.

These silicides were recently studied in detail by Austrian investigators [468–471]. The phases obtained, V_5Si_3, Nb_5Si_3, Ta_5Si_3, and those similar to them, all had a $D8_8$ type of structure. Together with Nb_5Si_3, phases with the formulas Nb_3Si_2 and Ta_3Si_2 have been mentioned [470] and these result from metallographic investigations and analogies with the U–Si system [473]. However, it was then shown [471] on the basis of x-ray analysis that these phases should be considered as pertaining to Me_5Si_3 and isotypic not only within the limits of transition metals of Group V, but also of Group VI. These phases with a $D8_8$ structure are especially stabilized by traces of carbon, nitrogen, and oxygen [474, 475, 634]. In addition, two tetragonal phases of the same composition were found [472]. According to these new data, α-Nb_2Si and β-Nb_2Si of Brauer and Scheele belong to an Nb_5Si_3 phase of different structure and composition (quite a large region of homogeneity). The similarity in

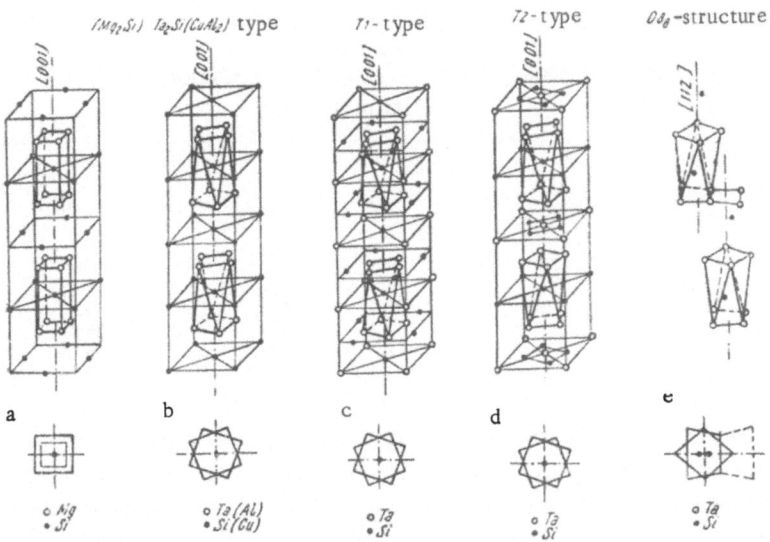

Fig. 75. Crystal lattices of different forms of the silicides Me_5Si_3 in comparison with Mg_2Si and Ta_2Si, according to Parthé, Lux, and Nowotny (1955): a) Mg_2Si, b) Ta_2Si, c) $T1$ type Me_5Si_3, d) $T2$ type Me_5Si_3, e) $D8_8$ type Me_5Si_3. Projections normal to the axis or the [112] direction are also given.

structure between $C16$ (Ta_2Si) and $T2$ types (Ta_5Si_3), which was demonstrated by French investigators [476], is based on the fact that they are all of the polyfluoride type (layers of the CaF_2 and $CuAl_2$ type). Figure 75 shows these structures, according to Parthé, Lux, and Nowotny [472], and Mg_2Si for comparison. The corners of the Archimedean antiprism are occupied by metal atoms and the internal points by silicon atoms. The angle of rotation of the basis to the upper plane in these structures is approximately 45°. With the $D8_8$ type, the antiprism is "destroyed," but the coordination obtained is found to be very stable.

The relation between the high- and low-temperature modifications of Me_5Si_3 phases has been studied very little as yet and therefore is not mentioned below. The Me_3Si_2 phases are replaced by Me_5Si_3 everywhere.

In the light of what has been said, the independent existence of Me_2Si phases for transition metals of Group V, especially Nb_2Si, requires special checking. The data in Table 22 show that the differences in the silicon contents of the Me_2Si, Me_5Si_3, and Me_3Si_2

Table 22

Silicon Content of Some Silicides of Groups V and VI
Transition Metals.

Metal	Silicon content of silicides, wt.%		
	Me_2Si	Me_5Si_3	Me_3Si_2
V	21.61	24.86	26.88
Nb	13.14	15.35	16.77
Ta	7.19	9.01	9.36
Cr	—	24.47	26.47
Mo	—	14.94	16.33
W	—	8.39	9.24

phases are not so great that they may be the sole criteria in metallographic analysis, making it possible to identify accurately the composition of these silicides, especially when one considers their wide and as yet incompletely determined regions of homogeneity. These remarks are also fully applicable to silicides of Group VI transition metals.

The silicides of Group V transition metals have even more strongly expressed metallic properties than titanium, zirconium, and hafnium silicides. The silicides of the Group V metals examined are also characterized by the fact that the maximal melting point is reached at a silicon content of about 40 at. %.

Apart from $Ta_{4.5}Si$, all the silicides of Group V transition metals are formed with a decrease in volume (see Table 2). The heats of formation of V_2Si, Ta_5Si_3, and $TaSi_2$ were determined by Robins and Jenkins [624].

V–Si SYSTEM. The first vanadium silicides were synthesized by Moissan and Halt [477] who reduced V_2O_3 and V_2O_5 with silicon by melting the mixture in an electric furnace. They also obtained vanadium silicides by the action of silicon on vanadium carbide at high temperatures or of copper silicide on vanadium. The reaction products were then treated with nitric or sulfuric acids and 10% KOH. These methods yielded vanadium silicides with the compositions V_2Si and VSi_2. These syntheses were later repeated by

Lebeau [478] and Baroduc-Müller [444]. Vanadium disilicide, VSi_2, was synthesized by Wallbaum [50] by heating a mixture of powdered vanadium and silicon in an alumina crucible in an argon atmosphere. These crucibles were also found to be suitable for fusion of vanadium disilicide. Cerwenka obtained VSi_2 by hot pressing this mixture at 1050° [27].

Giebelhausen [479] partly investigated the V–Si system, using vanadium of 94.2% purity. The silicon contained 0.7% Al. The alloys were prepared in quartz crucibles and contained 0–60% V. The investigations were carried out by means of thermal analysis and metallography. As a result of Giebelhausen's investigations, data were obtained on the structure of the phase diagram of the V–Si system in the VSi_2–Si section. It was found that vanadium disilicide does not form solid solutions with silicon. Solid solutions (between 47.5 and 60% V) were observed on the vanadium side of VSi_2.

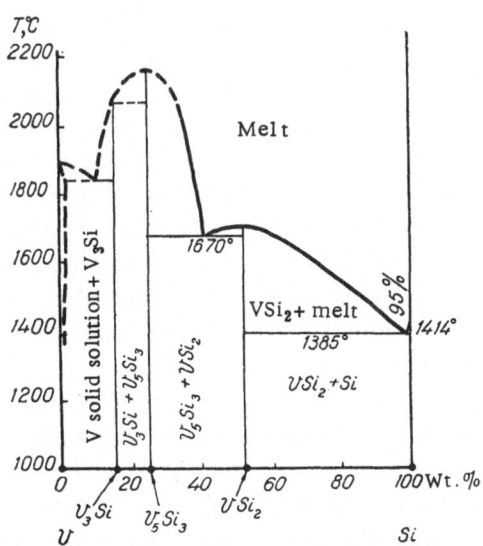

Fig. 76. Phase diagram of the V–Si system, according to Kieffer, Schmid, and Benesowsky (1955).

Vogel and Jenetzsch-Uschinski [480] investigated the Fe–V–Si system and found in the V–Si system silicides with the composition V_3Si, V_2Si, and VSi_2 and in the ternary system, $(Fe, Si)_4V_5$. Pearson [481] made a short survey of the V–Si system. According to Giebelhausen, the melting point of vanadium disilicide is 1655° and according to Cerenka, 1750°. A silicide with the composition V_5Si_3 was also synthesized.

The V–Si system was recently investigated by Kieffer, Schmid, and Benesowsky [482, 681]. They obtained samples for the investigation by hot-pressing or firing mixtures of pure vanadium and silicon in argon. The melting points were determined by a method similar to the determination of refractoriness. Silicon-rich alloys were also subjected to thermal analysis. All the preparations were investigated by microscopic and x-ray methods. As a result, a phase diagram was constructed for the V–Si system and this is shown in Fig. 76. Only three intermediate phases, V_3Si, V_5Si_3, and VSi_2, were found in it. The silicide with the composition V_2Si was not found. With very pure starting materials, the silicide V_5Si_3 was obtained as a tetragonal modification, but in the presence of even 0.13% C and 0.05% N, small amounts of this silicide were produced in a hexagonal form with a $D8_8$ type structure. With 0.25% C and 0.9% N content, only this hexagonal form of V_5Si_3 was obtained.

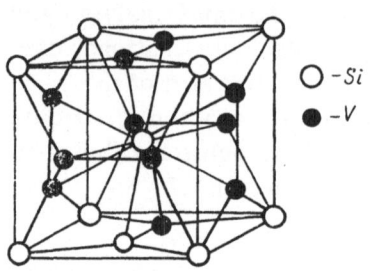

$\bigcirc$ –Si

$\bullet$ –V

Fig. 77. Crystal lattice of V_3Si (β-W type), space group O^3. Cr_3Si and Mo_3Si also crystallize in this type of lattice.

V_3Si melts incongruently and V_5Si_3 and VSi_2, congruently. Their melting points were found to be higher than Giebelhausen found, using less pure starting materials.

The results of investigation of the V–Si system presented agree in general with the preliminary experiments of Rostoker and Yamamota [638], who considered, however, that the VSi_2–Si eutectic melts at a considerably lower temperature than was established by Kieffer and his co-workers. However, it seems that the latter authors studied the V–Si system more fully and accurately than Rostoker and Yamamota.

The crystal structure of V_3Si, which was investigated by Wallbaum [483], was found to be cubic of the β-W type (Fig. 77). The silicides Cr_3Si and Mo_3Si crystallize similarly. The structures of vanadium, niobium, tantalum, and rhenium disilicides were also determined by Wallbaum [484]. Vanadium disilicide has a hexagonal structure, built up from layers of atoms with the densest packing.

The structures of the two V_5Si_3 modifications were determined by Parthé and his co-workers [472]. The characteristics of the elementary cells of vanadium silicides are presented in Table 2.

Alloys of vanadium with silicon are very friable. The hardness of vanadium disilicide is comparatively low (see Table 2). The microhardness of vanadium silicides equals 500–1200 kg/mm^2 [482], i.e., low.

According to Hönigschmid [23], V_2Si has a strong metallic luster. Fluorine readily decomposes this silicide on slight heating and chlorine and bromine do so at red heat. Hydrogen chloride reacts with it at 800° with the formation of a binary chloride of vanadium and silicon. Heating with carbon in an electric furnace partially decomposes V_2Si with the formation of carbide. With the exception of HF, mineral acids do not decompose it, while molten alkalies do. However, the individuality of this silicide still requires checking (see above).

Vanadium disilicide, VSi_2, crystallizes in the form of prisms with a metallic luster. Fluorine and chlorine decompose it at red heat with the formation of halides of both components. Oxygen, sulfur, and H_2S react slowly with vanadium disilicide only at bright red heat. Mineral acids and alkali solutions do not act on it. Even very dilute hydrofluoric acid dissolves this compound rapidly. Molten alkalies convert it to soluble vanadates and silicates. Hydrogen chloride reacts with vanadium disilicide on heating to form vanadium and silicon chlorides.

The chemical properties of other vanadium silicides have been studied little. The resistance of alloys from the V–Si system to oxidation at high temperatures is low, especially at vanadium contents greater than 33 at. %.

Vanadium silicides have no practical application. The comparatively low hardness and quite considerable chemical reactivity of vanadium silicides reduce their value in technology. The very low electrical resistance of vanadium disilicide (see Table 2) may be of use in the production of contacts for conducting ceramics.

Nb–Si SYSTEM. The first niobium silicide ($NbSi_2$) was synthesized by Wallbaum [484] by sintering a mixture of silicon and niobium in an alumina crucible in an argon atmosphere. Brauer and Scheele [27] repeated this synthesis at 1500–1700°. The action of $SiCl_4$ vapor on metallic niobium in the presence of hydrogen at 1100–1800° yielded $NbSi_2$.

Brauer and Scheele found that the solubility of silicon in solid niobium is less than 5 at. %. According to these authors, in addition to $NbSi_2$, two modifications of Nb_2Si exist. Silicides with the composition Nb_5Si_3 were synthesized and studied by Parthé and his co-workers [470, 472]. The phase diagram of the Nb–Si system was worked out by Knapton [474]. The alloys were prepared from

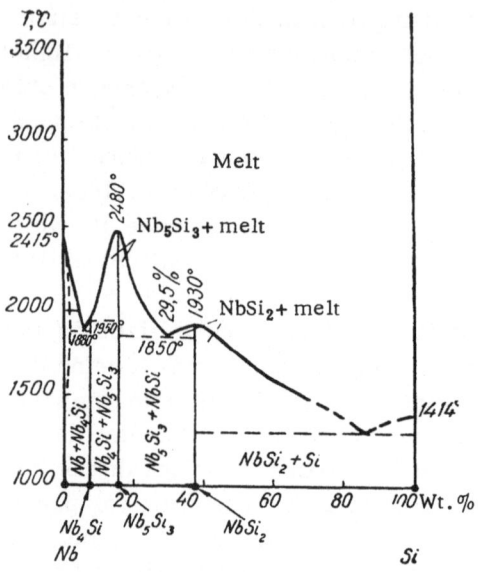

Fig. 78. Phase diagram of the Nb–Si system, according to Knapton (1955).

pure materials. They were investigated by x-ray and metallographic methods. The results of the investigations are shown in Fig. 78. The presence of the three following intermediate phases was established: Nb_4Si (isomorphous with the analogous compound of tantalum), Nb_5Si_3 (two tetragonal modifications with a transition point between 1900 and 2100°), and $NbSi_2$. The normal form of Ta_5Si_3 is isomorphous with the low-temperature modification of Nb_5Si_3 and the normal forms of V_5Si_3, Cr_5Si_3, Mo_5Si_3, and W_5Si_3 are isomorphous with the high-temperature modification of Nb_5Si_3. The solubility of silicon in niobium and niobium in silicon were not determined by Knapton, though silicon may be soluble in niobium, judging by the increase in the edge of the niobium cell from 3.2993 to 3.3083 kX at 1300° in the presence of silicon. The Si–$NbSi_2$ eutectic melts [681] at 1405° and contains 8% Nb.

Kieffer and his co-workers [482] also investigated the Nb–Si system and did not find the silicide Nb_4Si, though we do not consider this reliable, bearing in mind the existence of phases with high metal contents in the V–Si and Ta–Si systems.

Apparently, the silicide Nb_2Si does not exist. As was shown by Parthé [472], the silicide Nb_5Si_3 may have not two, but three structural forms (one in the presence of the stabilizers carbon, nitrogen, and oxygen).

The crystal structures of Nb_5Si_3 were studied by Parthé [472] and that of $NbSi_2$ by Wallbaum [484]. They are isomorphous with the corresponding vanadium silicides. The characteristics of the elementary cells and other properties of niobium silicides are presented in Table 2.

Niobium disilicide has some plasticity at high temperatures [516].

Its chemical properties are similar to those of VSi_2. The silicide Nb_5Si_3 is quite a reactive substance. When fused with TaC, Mo_2C, and Mo_2B, it forms three, and with TiB_2, one intermediate phase of as yet undetermined composition [660]. The hardness of niobium silicides is in the range $500-900$ kg/mm^2 (on the Vickers scale). They are very friable and oxidize quite strongly in air at high temperatures.

Niobium silicides have been studied little and are not yet used in technology.

Ta–Si SYSTEM. Tantalum disilicide, $TaSi_2$, was first synthesized by Hönigschmid [486] by aluminothermy from a mixture of Ta_2O_5, SiO_2, aluminum, and sulfur. Wallbaum [484] obtained this compound by sintering tantalum and silicon in an alumina crucible in an argon atmosphere. Brewer and his co-workers [487] obtained tantalum silicides with the compositions Ta_5Si, Ta_5Si_2, and Ta_5Si_3 in molybdenum or tantalum crucibles at $1610-2100°$ (mainly at $1900°$) and also determined a series of melting points in the Ta–Si system. Tantalum disilicide was obtained even at $1250°$ by hot-pressing of a mixture of the elements.

The Ta–Si system has been studied by Austrian investigators [468, 488].

The starting materials were very pure tantalum and silicon of 99% purity. The materials were ground in trichloroethylene. The preparations were obtained by hot-pressing [490] and then annealed at $1800°$ or fused in ThO_2 crucibles at $1950-2300°$. By x-ray investigation of the preparations obtained [489] it was shown that the Ta–Si system contained such intermediate phases as $Ta_{4.5}Si$, Ta_2Si, Ta_5Si_3 (several modifications with transition points at 1600 – 1800°), and $TaSi_2$. The properties of these phases are presented in Table 2. They are all isomorphous with the corresponding niobium silicides.

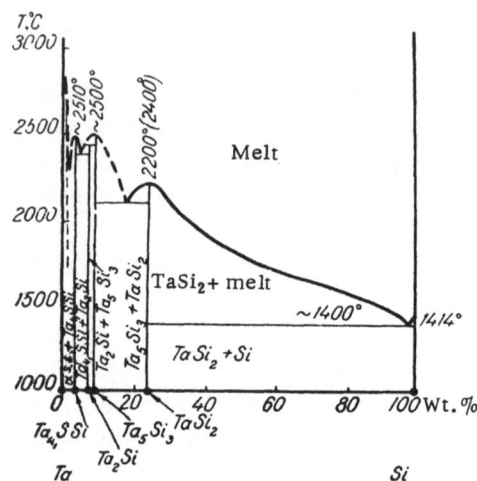

Fig. 79. Phase diagram of the Ta–Si system, according to Kieffer, Benesowsky, Nowotny, and Schachner (1953).

Tantalum dissolves 0.2% of Si at 1800° with expansion of the lattice. Silicon hardly dissolves tantalum.

The $Ta_{4.5}Si$ phase may be regarded as Ta_3 ($Ta_{0.28}Si_{0.72}$), with a structure of the Ni_3Ti type, but with the c axis doubled. The maximal melting point corresponds to the composition Ta_4Si. The region of homogeneity of this phase is approximately limited by the compositions Ta_5Si and Ta_4Si. There is no phase with the composition Ta_3Si.

The silicide Ta_2Si has a very narrow region of homogeneity. This silicide, which crystallizes like $CuAl_2$ (see Fig. 71), is isomorphous with the boride Ta_2B.

Three forms were found for the phase Ta_5Si_3 [472]. The structure of $TaSi_2$ was determined by Wallbaum [484].

Figure 79 shows that tantalum silicides have very high melting points. These compounds are the most refractory of all the silicides. Only hafnium silicides should be close to them in melting point. Titanium silicides are quite resistant to oxidation when heated in air as is shown by the change in weight after being heated at 1500° in air for one hour (Fig. 80). Alloys close to the disilicide in composition (about 70 wt.% Ta) are particularly resistant to oxidation and at high temperatures, these are covered with a glassy film of SiO_2 formed by oxidation. This film also protects the alloy from further oxidation.

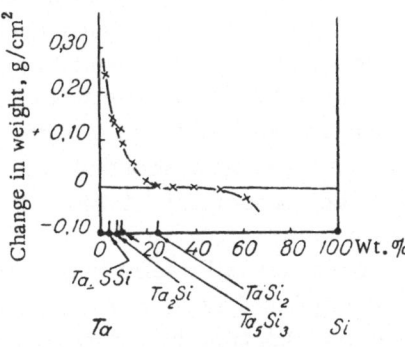

This property of tantalum silicides is of great practical interest from the point of view of obtaining protective coverings on tantalum. Alloys of tantalum and silicon are very friable, but tantalum disilicide shows plasticity at high temperatures [516].

Fig. 80. Change in weight of alloys from the Ta—Si system after being heated in air at 1500° for one hour, according to Kieffer, Benesowsky, Nowotny, and Schachner (1953).

Brewer and his co-workers [487] found somewhat lower melting points for the eutectics than those presented according to Kieffer's data [488] and these were:

Eutectic between the phases	Melting point, °C
Ta–Ta$_5$Si	2110
Ta$_5$Si–"Ta$_5$Si$_2$"	1910
"Ta$_5$Si$_2$"–Ta$_5$Si$_3$	1610
Ta$_5$Si$_3$–TaSi$_2$	1770

According to Cerwenka, the melting point of tantalum disilicide is 2400° [27]. The determinations of Kieffer are apparently more accurate than those of Brewer.

Tantalum disilicide crystallizes in the form of beautiful blue-gray prisms with a metallic luster and with pyramidal ends. This compound is very stable chemically. Its resistance to oxidation when heated in air was mentioned above (Fig. 80). Fluorine reacts with this disilicide on heating and chlorine, only at high temperatures. Tantalum disilicide is oxidized when heated in oxygen. Among the mineral acids, only hydrofluoric acid acts on it. Molten alkalies readily decompose this compound [23].

When tantalum disilicide is fused with TiC, TiB$_2$, or ZrB$_2$, no chemical interaction occurs, but with TaC, Mo$_2$C, or Mo$_2$B, compounds of as yet unknown composition are formed.

The silicide Ta$_5$Si$_3$ is more reactive than TaSi$_2$. Fusion of Ta$_5$Si$_3$ with TiB$_2$ yields one, with TaC two, and with Mo$_2$C or Mo$_2$B, three compounds of undetermined composition [660].

According to Gallistl, the electrical resistance of hot-pressed tantalum disilicide is about 8.5 μohm·cm [27]. This compound passes into a superconductive state at 4.4° K [491].

The properties of the other tantalum silicides have not been described.

No practical use is known at the moment for tantalum silicides. The high melting points and the resistance to oxidation of some of them on heating in air indicate the need for their investigation as refractories for very special purposes. Restrictions arise here due to the high cost of tantalum and the possibility of its use in the metallic form as a high temperature material.

Systems Formed by Silicon with Transition Metals of Group VI

In the section on systems formed by silicon with transition metals of Group V the reasons were given for the similarity between the systems and those formed with transition metals of Group VI. In the same place we mentioned the latest investigations, which showed that the previously established silicides Me$_3$Si$_2$ should be replaced by Me$_5$Si$_3$ [471]. The greatest difficulties here occurred with the W–Si system, since the hexagonal phase W$_5$Si$_3$ could not be stabilized by carbon [469]. However, W$_5$Si$_3$ was later obtained in the form of a monocrystal [492] of the tetragonal system. It was therefore necessary to introduce corrections into the phase diagrams of the Cr–Si, Mo–Si, and W–Si systems and replace the silicides Me$_3$Si$_2$ by the compounds Me$_5$Si$_2$, which are similar to them in composition (Table 22), and this we did. There is no doubt that the Me$_5$Si$_3$ phases have a much greater region of homogeneity than other silicides of the systems examined here. However, the limits of the regions of homogeneity of all the silicides of transition metals of Groups IV, V, and VI, including those of the type Me$_5$Si$_3$, have not been established yet. Therefore, also the phase diagrams of these systems which are described must be regarded as preliminary to a considerable extent.

The silicide W_3Si has not been isolated yet for the W–Si system. though there can be hardly any doubt as to its existence, considering the existence of the compounds Me_3Si in the Cr–Si, Mo–Si, and Re–Si systems and also of the silicide $Ta_{4.5}Si$, corresponding to Ta_3 ($Ta_{0.28}Si_{0.72}$). It must be assumed that the existence of the silicide W_3Si will soon be established.

The interesting properties of tungsten and especially molybdenum disilicides were the reason for investigating in greater detail the properties of these silicides and for considering the problem of their use [617, 618]. Therefore more attention is paid to them below than to chromium silicides.

Silicides of Group VI transition metals have a metallic appearance. They are grayish in color. They are formed with a decrease in volume (see Table 2). They are all very friable like the silicides of Group IV and V transition metals. Details of their properties that have been studied are presented in the descriptions of the separate systems.

With a decrease in the ratio of the atomic radii of the Group IV, V, or VI transition metals and silicon, the disilicides show a change from the tetragonal to the hexagonal system. However, the known structures of molybdenum and tungsten disilicides belong to the tetragonal system, though this ratio (1.20 and 1.21) is within the range where hexagonal disilicides are found (chromium – 1.16 and vanadium – 1.35). Therefore it must be assumed that molybdenum and tungsten disilicides must also exist as hexagonal modifications, which have not been discovered yet. It is possible that these modifications are stable only at temperatures above 1000° and therefore have not been found yet. Further investigations, especially at high temperatures, should solve this problem.

Cr–Si SYSTEM. Zettel (1897) obtained the silicide Cr_3Si by fusing a mixture of copper, aluminum, and Cr_2O_3 in a fire-clay crucible. Warren (1898) also obtained chromium silicides. Lebeau [493] obtained this silicide by melting a mixture of chromium, copper, and silicon in an electric furnace. Moissan (1895) considered that he obtained a silicide with the composition Cr_2Si by heating a mixture of chromium and silicon or by reducing Cr_2O_3 and SiO_2 with carbon.

Lebeau [493] and Vigouroux [494] also described a silicide with the composition Cr_3Si_2. Chromium disilicide, $CrSi_2$, was obtained by De Chalmot [495] by heating the components. Later it was also obtained by Lebeau [493]. Frilley [496] prepared 19 alloys with Si contents of from 10 to 89% and determined their specific gravities. He found that the melting points of these alloys were lower than those of chromium and silicon. Frilley considered that the chromium silicides Cr_3Si, $CrSi$, and $CrSi_2$ exist. Breaks on the specific gravity curve were observed at compositions corresponding to the formulas Cr_3Si, Cr_2Si, $CrSi$, Cr_2Si_3, $CrSi_2$, and Cr_2Si_7.

By x-ray investigation, Boren [497] established that four silicides exist in the Cr–Si system: cubic Cr_3Si; a phase of unknown composition, stable at temperatures below 1000°; cubic $CrSi$ (FeS type) and hexagonal $CrSi_2$. It was found that chromium in the solid state dissolves up to 1% Si, while silicon does not dissolve chromium. Cerwenka (1951) obtained chromium disilicide by hot-pressing of a mixture of the components at 1050°.

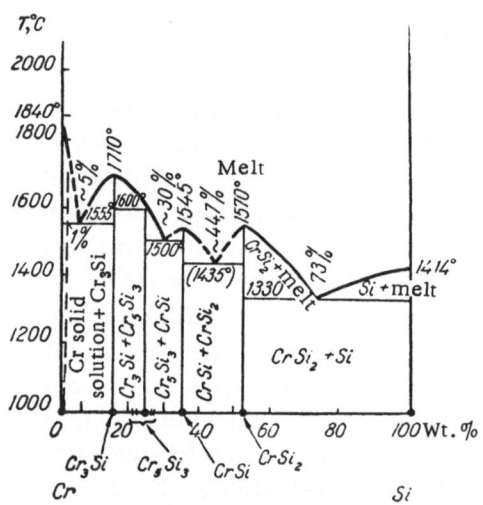

Fig. 81. Phase diagram of the Cr–Si system according to Kieffer, Benesowsky and Schroth (1953). The silicide Cr_3Si_2 is replaced by Cr_5Si_3, according to the latest data.

An investigation of the phase diagram of the Cr–Si system was undertaken by Kurnakov [498], who found that all four chromium silicides melt congruently and at the following temperatures: Cr_3Si–1750°, Cr_2Si (flat maximum) –1606°, $CrSi$–1630°, $CrSi_2$–1500°. This system was investigated more fully by Kieffer, Benesowsky, and Schroth [499], who established that the silicide "Cr_3Si_2" melts incongruently. The rest of the data they obtained were similar to Kurnakov's diagram. It should be noted that silicon used in Kieffer's experiments contained about 1.3% Fe. Therefore, the melting point of the silicon-rich silicide $CrSi_2$ was found to be 20° lower than the value determined by Cerwenka [26].

It was pointed out above that the silicide "Cr_3Si_2" corresponds more accurately to the formula Cr_5Si_3. Kieffer [499] reported that this phase is apparently unstable at temperatures above 856°, but this was not reflected in the phase diagram he proposed for the Cr–Si system. Its region of homogeneity extends approximately to the composition Cr_2Si.

Considering the circumstances given and also the results of a new determination of the melting point of chromium [500], in Fig. 81 we give a corrected phase diagram of the Cr–Si system, which requires no particular explanations. The properties of chromium silicides are presented in Table 2.

The silicide Cr_3Si forms prismatic crystals which cut glass but not quartz [23]. The structure of this silicide is the same as that of V_3Si. Chlorine and bromine react with it at red heat, while no reaction is observed with sulfur. The compound Cr_3Si does not dissolve in HCl

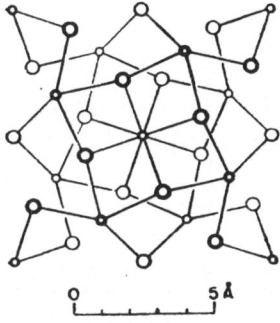

Fig. 82. Structure of Cr_5Si_3 projected onto the (001) plane, according to Dauben, Templeton, and Myers (1956). The lines show distances less than 3.50 A. Cr_5Si_3 has a tetragonal cell, D_{4h}^{18} — $I4/mcm$; a = 9.170 A, c = 4.636 A, $z = 4$. The small circles represent Si atoms and the large ones, Cr.

or HNO_3, but dissolves readily in HF, even at room temperature. Molten KCl and K_2CO_3 react with it slowly, but it dissolves more readily in molten alkalies or mixtures of alkali nitrates and carbonates.

The silicide "Cr_3Si_2" (Cr_5Si_3) crystallizes in the form of long tetragonal prisms, which cut glass but not quartz. The hardness increases with a change to the composition "CrSi."

The structure of the silicide with this composition was determined by Parthé [470], who found the following dimensions for the elementary cell: $a = 9.16$ kX, $c = 4.64$ kX, $z = 6$, but it was later established that Cr_5Si_3 has a tetragonal structure with the elementary cell dimensions presented in Table 2. This silicide is apparently dimorphous. Its hexagonal form is of the $D8_8$ type and is stabilized by traces of carbon, nitrogen, or oxygen. The structure of its tetragonal form was recently studied in more detail [639]. There are no Si–Si bonds in it. A projection of the Cr_5Si_3 structure is presented in Fig. 82.

The electroconductivity and thermoelectric properties of chromium silicides were studied by Guseva and Ovechkin [682].

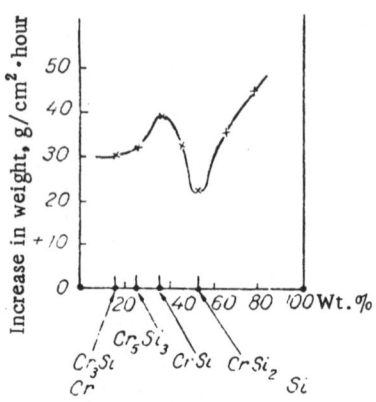

Fig. 83. Resistance of alloys from the Cr—Si system to oxidation during heating in air (1300°), according to Kieffer, Benesowsky, and Schroth (1953).

Chlorine reacts with this silicide at 400°, bromine reacts only at cherry red heat [23]; iodine does not react with it at all. Gaseous HCl decomposes "Cr_3Si_2." In contrast to the dilute acid, concentrated hydrochloric acid dissolves it. This silicide is resistant to the action of HNO_3 and H_2SO_4, but dissolves very readily in hydrofluoric acid. Molten KNO_3 and KCl do not react with "Cr_3Si_2," while molten alkali carbonates readily convert it to alkali silicates and chromic oxide.

The chemical properties of chromium monosilicide, CrSi, have been studied little. Its structure, which is analogous to FeSi, is characterized by the presence of heptagons of silicon atoms with chromium atoms at their centers.

Chromium disilicide, $CrSi_2$, forms needlelike crystals, isomorphous with niobium and tantalum disilicides, which do not react in the cold with hydrochloric acid or aqua regia, but dissolve in hydrofluoric acid [23]. Chromium disilicide reacts with boron at 1550° [27] to form CrB.

The resistance of alloys from the Cr–Si system to oxidation during heating in air is shown by Fig. 83 [499]. In contrast to molybdenum and tungsten silicides, chromium silicides do not form compact, gas-resistant films when heated in air. The films formed on alloys of this system are not heat-resistant and are deformed on heating. This is explained by the formation of SiO_2 and Cr_2O_3, which do not give glasslike films. Chromium oxide is involatile and difficult to sinter. This also explains the comparatively high oxidizability of Cr–Si alloys when heated in air.

On the basis of preliminary data of Mondolfo, Pratt, and Rainer, Robinson [501] established that partial replacement of the silicon atoms and even chromium atoms by aluminum is possible in chromium disilicide. As a result, the so-called β-(AlCrSi) is obtained and this belongs to the same space group as chromium disilicide (D_6^4) and has the cell dimensions: $a = 4.49$ A, $c = 6.377$ A, i.e., somewhat larger than for chromium disilicide (see Table 2). This phase is a solid solution with substitution of aluminum in chromium disilicide, which is present in some aluminum alloys.

As Robinson showed, the phase called α-(AlCrSi) by Mondolfo has a composition coresponding to the formula $Cr_4Si_4Al_{13}$ and is found in ternary Al–Cr–Si alloys. The lattice of this electronic compound is cubic with $a = 10.917 \pm 0.001$ A, $z = 4$, the space group T_d^2 and a specific gravity of 3.405. This compound is similar to Mn_3SiAl_9.

The ternary system Cr–Si–C is similar to the Mo–Si–C system, which is described below [470]. Here there is a ternary phase with the composition $Cr_{5-x}Si_{3-y}C_{x+y}$, of the $D8_8$ type.

Chromium silicides are not yet of independent practical value. Due to their quite high oxidizability and comparatively low melting points, these compounds can hardly find application in refractory technology. The Cr–Si system is of interest primarily for ternary and polycomponent alloys (Cromansil steel, containing approximately up to 1% Cr, Si, and Mn, which is used in ship and aircraft construction, in the building of bridge girders, etc.).

Mo–Si SYSTEM. Moissan [502] used an electric furnace to prepare molybdenum silicides which did not melt when oxyhydrogen gas was injected. Vigouroux [503] obtained these compounds by reduction of molybdenum oxides with silicon, also in an electric furnace. Molybdenum disilicide, $MoSi_2$, was first obtained in a pure form by Hönigschmid [486] by aluminothermy, starting from a mixture of 10 g of MoO_3, 90 g of SiO_2, 100 g of Al, and 125 g of S. This mixture was fused in a chamotte crucible and then treated with HCl and an alkali solution, as a result of which, pure molybdenum disilicide was isolated. Defacqz [504] obtained it by using copper silicides and molybdenum.

Vigouroux [503] considered that he also obtained a silicide with the composition $MoSi_3$, but he did not present convincing data on the individuality of this compound. Brewer and his co-workers [487] obtained molybdenum silicides by sintering a mixture of molybdenum and silicon in an argon atmosphere at 1660–2020° and Wedekind (German Pat. No. 294267, 1913 [27]) obtained them by hot-pressing of this mixture. Brewer obtained silicides with the compositions $MoSi_2$, Mo_3Si, and $MoSi_{0.65}$ ($Mo_{20}Si_{13}$, which is very close to Mo_5Si_3).

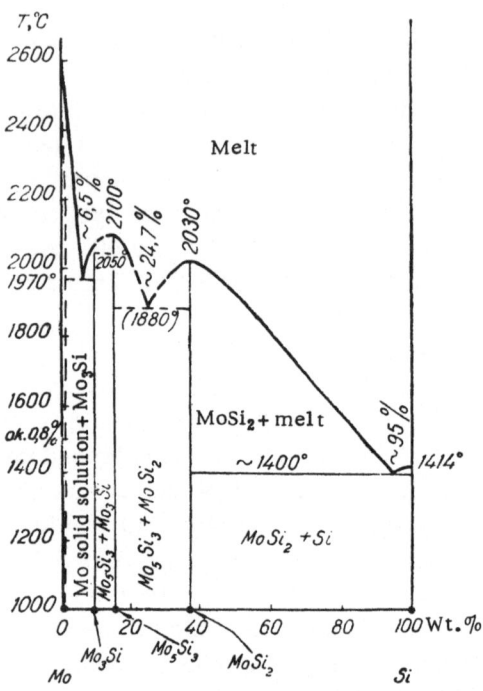

Fig. 84. Phase diagram of the Mo–Si system, according to Kieffer and Cerwenka (1952), with corrections by Nowotny, Kieffer, Parthé, and Benesowsky (1954). The silicide "Mo_3Si_2" is replaced by Mo_5Si_3, according to the latest data.

Wedekind considered that he obtained the silicide MoSi, which, like Mo_2Si_3, was not found in later investigations. Campbell and his co-workers [516] obtained molybdenum silicides in the form of a layer on metallic molybdenum by the action of $SiCl_4$ vapor and H_2.

Ham [505] established that molybdenum dissolves 0.8% of silicon at 1430° and 0.15% at 1200°. Brewer [487] found that the eutectics in the Mo–Si system melt at the following temperatures:

$$Mo–Mo_3Si \dots \dots \dots \quad 2160°$$
$$Mo_3Si–MoSi_{0.65} \dots \dots \quad 1850°$$
$$MoSi_{0.65}–MoSi_2 \dots \dots \quad 1850°$$

The phase diagram of the Mo–Si system was worked out by Kieffer and Cerwenka [506] and was then corrected in the Mo_3Si part by Nowotny, Parthé, Kieffer, and Benesowsky [153]. It was then shown [469, 492] that the silicide Mo_3Si_2 corresponds more accurately to the formula Mo_5Si_3. The phase diagram of the Mo–Si system, allowing for all these refinements, is presented in Fig. 84 and does not require further explanations.

According to the determinations of American investigators [27], $MoSi_2$ melts at 1870° with decomposition in the presence of carbon. Thus, the melting point of $MoSi_2$ is reported differently by different authors and therefore requires more accurate determination.

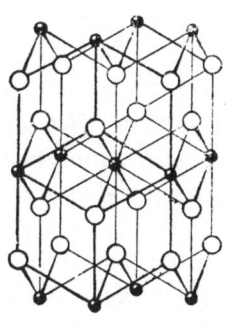

⬤ Mo ◯ Si

Fig. 85. Crystal lattice of $MoSi_2$. Space group $D_{4h}^{17} - I4mmm$, $z = 2$. WSi_2 and $ReSi_2$ also crystallize in this way.

In an investigation on the Mo–Si system, it was shown that molybdenum silicides are formed with a strong exothermal effect. Etching with HF by heating for 0.5 min or without heating with a mixture of HF and HNO_3 for five sec was recommended for studying polished sections of alloys from this system. It was difficult to find a crucible material for melting "Mo_3Si_2" which did not react with the melt. This compound may undergo conversions at temperatures below 1000°.

The crystal structure of Mo_3Si was found to be the same as that of V_3Si [507]. The characteristics of its elementary cell and those of other molybdenum silicides are presented in Table 2. In Mo_3Si, the silicon atoms are surrounded by 12 molybdenum atoms at a distance of 2.73 A. Each molybdenum atom is surrounded by two molybdenum atoms (at a distance of 2.44 A), four silicon atoms (2.73 A), and eight molybdenum atoms (2.99 A). Thus, this structure is similar to that of β-W.

If it is assumed that the composition of the next silicide corresponds to Mo_3Si_3, then its lattice is tetragonal [470], $a = 9.64$ kX, $c = 4.98$ kX, $z = 6$. However, the composition of this compound and the lattice characteristics presented in Table 2 are more firmly established [469, 470, 492]. The structure of the $D8_8$ type is stabilized by the presence of carbon, nitrogen, or oxygen, but for the pure silicide a $T1$ structure is stable, i.e., isomorphous with the analogous silicides of niobium and tantalum.

The structure of molybdenum disilicide was investigated by Zachariasen [508]. It is illustrated by Fig. 85. Here the silicon atoms form a framework with the molybdenum atoms arranged in the spaces. Figure 85 shows that molybdenum disilicide has a layer structure, consisting of two layers of silicon atoms and one layer of molybdenum atoms. Such a structure causes molybdenum disilicide to have some plasticity at high temperatures (above 1600°) [509, 517], and in this, the compound has an advantage over the highly refractory carbides, nitrides, and borides. At room temperature molybdenum disilicide is very friable and has a low impact strength – less than 0.1 kgm [652]. It is plastic at high temperatures and can be hot-pressed [654]. The microhardness of this compound is comparatively low (see Table 2). The compression strength of hot-pressed molybdenum disilicide samples is 4000–6000 kg/cm^2 and sometimes up to 7000 kg/cm^2 [663]. The bending strength is 4780 kg/cm^2 at 980°, 5970 at 1100°, and 3790 at 1200°. In individual cases, the bending strength of articles from this compound falls to 2500 kg/cm^2.

Ault and Deutsch [655] studied the creep of hot-pressed molybdenum disilicide samples at 870–1095° with loads of 700–1760 kg/cm^2. The time under load was 100–300 hours. It was established that the creep of this disilicide is high at temperatures above 1000°. For example, the elongation was about 10% after 100 hours at 1040° under a load of 1000 kg/cm^2. With the same load period, the strength was 2100 kg/cm^2 at 980°, 1050 at 1040°, and 630 at 1100°.

Under a load of 15 kg/cm^2, molybdenum disilicide samples began to soften at 1680–1700° with a continuously increasing temperature [652]. Additives usually reduced the softening point of samples under load sharply (sometimes by several hundred degrees). The heat resistance of molybdenum disilicide articles during water cooling is a factor of approximately two less than that of BeO articles.

Molybdenum disilicide was shown to have metallic properties by a study of the bonds in its lattice [510] and the changes in electrical resistance with temperature [511, 517]. The data on electrical resistance presented in Table 23 was obtained for a hot-pressed sample of molybdenum disilicide (porosity about 2.5%).

Table 23

Electrical Resistance of Molybdenum Disilicide at Various Temperatures

Temperature, °C	Electrical resistance, μ ohm $\cdot$ cm	Increase in electrical resistance, % per °C
—80 [511]	18.9	—
+22 [511]	21.5	0.135
+65 [511]	22.7	0.129
1600 [517]	75—80	—

The temperature coefficient of electrical resistance of molybdenum disilicide is positive, but decreases with temperature. The absolute value of its electrical resistance at room temperature corresponds approximately to that of low-carbon silicon steel.

The thermoelectric effect in molybdenum disilicide relative to platinum was measured [512] at from 210 to 873° K. The thermoelectromotive force, E, in microvolts, for these conditions, relative to temperature t (°C) is expressed by the equation

$$E = 5.13t + \frac{3.76 \cdot 10^{-2}}{2} t^2 - \frac{15.65 \cdot 10^{-6}}{3} t^3.$$

Hence, values close to those for copper and platinum are obtained and this also indicates that $MoSi_2$ has metallic properties.

The thermodynamic properties of molybdenum disilicide were studied by Douglas and Logan [513] under a US Air Force contract on a sample of 98.2% purity. It was thus established that molybdenum disilicide does not react with a Nichrome lining up to 1000°. The measurements were carried out in a helium atmosphere up to 900° and corrected for impurities. The results obtained are expressed by the following equations:

$$H°_T - H°_{298.16} = 16.944 \cdot T + 0.001304 \cdot T^2 - 1519.2 \lg T - 1409$$

$$(\text{accuracy} \pm 0.05\%),$$

$$S°_T - S°_{298.16} = 39.016 \lg T + 0.00261 \cdot T + \frac{660}{T} - 99.53,$$

$$C°_p = 16.94 + 0.00261 \cdot T - \frac{660}{T},$$

where T is the absolute temperature.

No anomalies in C_p (T) were found up to $T = 1200°$ K.

Walker and his co-workers [514] gave the following equations for the heat content and the heat capacity of molybdenum disilicide:

$$H_T - H_{303.16} = 0.1332\,T - 21.83\lg T - 112.2649\,(T = 303 - 598°\,\text{K}),$$

$$H_T - H_{303.16} = 0.1404\,T - 31.93\lg T + 11.8283\,(T = 598 - 1150°\,\text{K}),$$

$$C_p = 0.1332 - \frac{9.477}{T}\ (T = 303 - 598°\,\text{K}),$$

$$C_p = 0.1404 - \frac{13.864}{T}\ (T = 598 - 1150°\,\text{K}).$$

The data presented are of importance not only in calculating the heat consumed in heating molybdenum disilicide articles, but also in thermodynamic calculations of the possibility of reactions of various substances with this silicide on heating.

The heat of formation of molybdenum disilicide, ΔH_{298}, was determined by Robins and Jenkins [624] and also by Brewer and Krikorian [689] and found to be as follows: according to Robins and Jenkins, −16 kcal/mole, according to Brewer and Krikorian, −8 to −18 kcal/mole.

If a brick of a compressed mixture of powdered molybdenum and silicon in the ratio $MoSi_2$ is heated at one end, then the reaction occurs with so much heat evolution that it is sufficient to produce reaction through the rest of the material. The whole of the brick is thus converted into molybdenum disilicide.

Table 24

Low-Temperature Oxidation Resistance of Molybdenum−
Silicon Alloys, Obtained by Powder Metallurgy,
in Relation to Composition [520]

Composition, wt.%		Atomic composition	Sintering temper- ature, °C	Phases found in alloy by x-ray methods	Increase in wt. at 600° in a stream of air, g/m² ·hour
Mo	Si				
91.1	8.9	Mo_3Si	1800	Mo_3Si	300—600
87.2	12.8	Mo_2Si	1800	$Mo_3Si + Mo_3Si_2$	500—900
77.4	22.6	MoS	1800	$Mo_3Si_2 + MoSi_2$	8—10
69.6	30.4	Mo_2Si_3	1800	$MoSi_2 + Mo_3Si_2$	20
66.4	33.6	Mo_3Si_5	1800	$MoSi_2$	5—6
63.2	36.8	$MoSi_2$	1450	$MoSi_2$	5—6

The coefficient of linear thermal expansion of a polycrystalline hot-pressed sample of molybdenum disilicide (porosity about 3.5%) equals $9.2 \cdot 10^{-6}$ (25−1500°) and the thermal conductivity is about 0.075 cal/cm·sec·deg (25−200°). The thermoelectric emission of this disilicide is very low. Articles from it have good heat resistance.

Table 25

High-Temperature Oxidation Resistance of Hot-Pressed
Molybdenum–Silicon Alloys in Relation to
Composition [490]

Composition, wt.%		Change in wt. at 1500° in a stream of air, g/m²·hour	Composition, wt.%		Change in wt. at 1500° in a stream of air, g/m²·hour
Mo	Si		Mo	Si	
90,9	9,1	—1804	69,5	30,5	+ 1
89,5	10,5	—1540	63,1	36,9	+ 3
88	12	— 140	60	40	+ 2
85	15	— 149	55	45	—11
83,7	16,3	— 141	50	50	—19
77,4	22,6	— 2			

Note. Date of Kieffer and Cerwenka, obtained over 4.5 hours and expressed in g/m², calculated as g/m²·hour.

Hönigschmid [23] noted that molybdenum disilicide, $MoSi_2$, has a considerable resistance to oxidation during heating in air or even in a stream of oxygen. Kieffer and Cerwenka [506] reported that the greatest resistance under these conditions was shown by alloys containing "Mo_3Si" and $MoSi_2$ (change in weight over 4.5 hours at 1500° from −81.5 to +2.81 g/m²) and the least resistance by Mo_3Si (Tables 24 and 25).

Table 26

Change in Weight of Molybdenum Disilicide
at Different Temperatures

Temperature, °C	Oxidation time, hours	Change in weight, g/cm²·hour	Author
1095	75	$-0.3 \cdot 10^{-6}$	Maxwell (1949)
1095	150	$-0.4 \cdot 10^{-6}$	Maxwell (1949)
1200	200	$+1.0 \cdot 10^{-6}$	Long (1950)
1200	300	$+0.7 \cdot 10^{-6}$	Long (1950)
1320	50	$+5.0 \cdot 10^{-6}$	Maxwell (1949)
1320	100	$+4.0 \cdot 10^{-6}$	Maxwell (1949)
1500	4.5	$+0.3 \cdot 10^{-3}$	Kieffer (1952)
1565	100	−3.67	Long (1950)
1565	135	−3.10	Long (1950)

Table 26, which was drawn up by Fitzer and Schwab [518], gives data on the change in weight of molybdenum disilicide samples in relation to temperature.

In addition to this, the characteristics of the oxidizability of molybdenum disilicide in a stream of oxygen, according to Fitzer [518] are shown in Figs. 86 and 87, from which it follows that this compound is very resistant. The reason for its resistance to oxi-

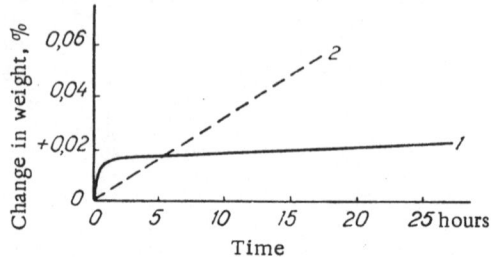

Fig. 86. Absorption of oxygen by MoSi$_2$ powder in relation to heating time: 1) at room temperature, 2) at 100°, according to Maxwell (1952).

dation is the formation of a thin protective layer of glassy silica. In contrast to oxides of Group IV and V transition metals and also chromium, MoO$_3$ gives a glass with silica, which has a low permeability to oxygen and therefore acts as a protective layer, preventing further oxidation of the molybdenum disilicide. As follows

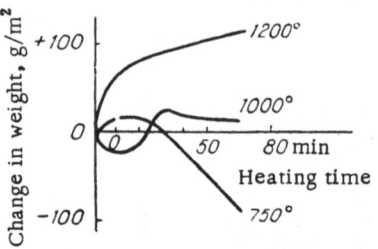

Fig. 87. Change in weight of sintered MoSi$_2$ in a stream of oxygen at various temperatures and short heating times, according to Fitzer and Schwab (1955).

from the change in weight of the samples tested (Fig. 87), this compound is oxidized up to the temperature at which a glassy film is formed (approximately up to 1300°).

At 400–600°, when no protective film is formed, even compact samples of molybdenum disilicide are readily oxidized in air and change into an amorphous powder in a few hours. This process has been called molybdenum disilicide "plague" in analogy with "tin plague." It should be noted, however, that this crumbling is the result of a chemical change in the case of molybdenum disilicide and a modification change with tin. Both cases are characterized by a sharp increase in volume, which naturally leads to crumbling. If the protective film of silica has already been formed on molybdenum disilicide, then this crumbling does not occur. This protective film is readily and rapidly formed on molybdenum disilicide above 1400° and is apparently a mixed oxide, containing not only SiO_2, but also molybdenum oxide. At temperatures of 700–1000°, a crystalline phase with a metallic luster and which transmits a red color in polarized light is formed on the surface of molybdenum disilicide, together with cristobalite [652]. The composition

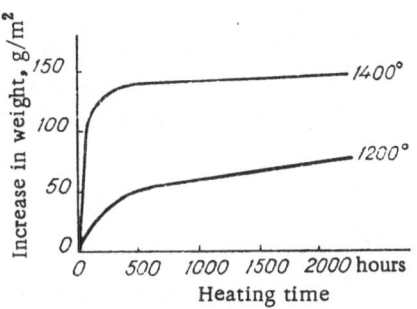

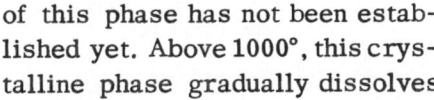

Fig. 88. Change in the weight of sintered $MoSi_2$ in a stream of oxygen at different temperatures and long heating periods, according to Fitzer and Schwab (1955).

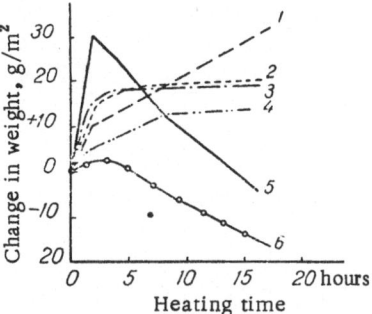

Fig. 89. Change in weight of $MoSi_2$ when treated with different gases with heating under identical conditions in relation to heating time, according to Kieffer, Benesowsky, and Konopicky (1954):   1) CO_2, 2) N_2,   3) combustion products of propane,   4) air,   5) CO,   6) mixture of propane and air.

of this phase has not been established yet. Above 1000°, this crystalline phase gradually dissolves in the silica to form a protective film, in which cristobalite cannot be detected. Above 1400°, this film on molybdenum disilicide is obtained in a strong and compact form which does not scale off on cooling. It remains in this way up to 1700°, but above this temperature it begins to melt and collect in drops. Therefore, articles of molybdenum disilicide cannot be used in air above 1700°.

Due to the absorption of oxygen by $MoSi_2$ powder even at room temperature (see Fig. 85), this material should be powdered in a readily removable protective liquid. $MoSi_2$ powder is pyrophoric and ignites at 500° [518].

Figure 88 shows that at high temperatures the weight of molybdenum disilicide samples is readily stabilized and then remains almost unchanged for long periods at 1400°. This phenomenon, together with the high electroconductivity, makes it possible in principle to make heaters from this material, like carborundum, and such heaters were made by Kieffer, Benesowsky, and Konopicky [517, 526] for operation at 1700° for several thousand hours. After 3000 hours, the thickness of the glassy film on this silicide, heated in a stream of oxygen, was 0.05–0.1 mm and after 2000 hours at 1400°, 0.02–0.03 mm [518].

Figures 89 and 90 show the changes in weight of sintered samples of molybdenum disilicide, heated in different gases. From the figures it follows that this silicide is resistant to the action of SO_2, NO, N_2, CO_2, and hydrocarbons. Strong corrosion occurs

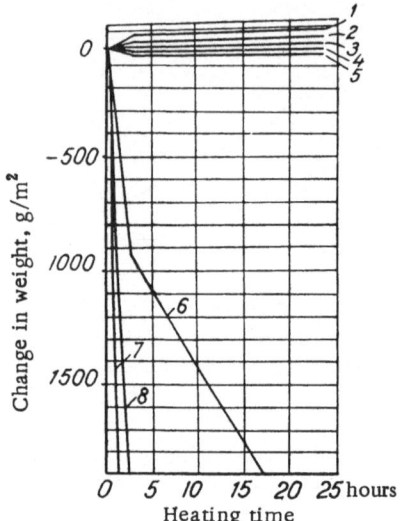

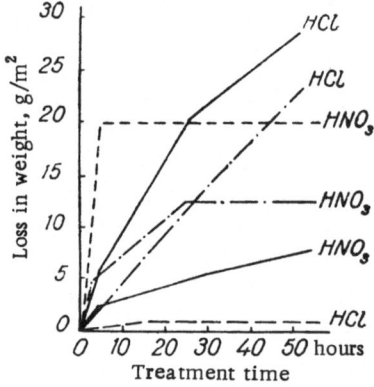

Fig. 90. Change in weight of $MoSi_2$ in various gases at 1000° in relation to the heating time, according to Fitzer and Schwab (1955): 1) CO_2, SO_2, uncovered sample, 2) HCl, covered sample, 3) CO_2, covered, 4) SO_2, covered, 5) NO, uncovered, 6) HCl, uncovered, 7) Cl_2, covered, 8) Cl_2, uncovered.

Fig. 91. Solubility of sintered $MoSi_2$ in boiling dilute (1:1) HNO_3 and HCl, according to Fitzer and Schwab (1955). Solid line—$MoSi_2$ without protective covering; broken line—$MoSi_2$ with protective covering obtained by preliminary heating in air; broken line with points—$MoSi_2$ with 5% Si added.

under the action of Cl_2 and HCl and is accompanied by the formation of volatile molybdenum and silicon chlorides. If samples of molybdenum disilicide have first been covered with a protective film by heating, then the resistance to gases increases, apart from the case of chlorine; the reason for the latter is still not known. Fluorine reacts with this compound even at room temperature, while bromine and iodine do not react even on heating [517]. Ammonia vapor is adsorbed by molybdenum disilicide (heat of adsorption more than 40 kcal/mole of adsorbed NH_3) and partially decomposes it with the formation of Si_3N_4 [651].

Molybdenum disilicide is insoluble in mineral acids (even in aqua regia or hydrofluoric acid [23, 26, 27, 517, 518, 685]) but rapidly dissolves in a mixture of HF and HNO_3 with the evolution of nitrogen oxides. The weak action of boiling dilute HNO_3 and HCl on molybdenum disilicide is illustrated by Fig. 91. A passivating layer is formed on the surface of this compound in nitric acid and this inhibits further solution, but passivation stops after 50 hours. The rate of solution in boiling HCl is about 15 g/m^2 per day [518]. The protective layer was found to be ineffective against the action of HNO_3 as was the addition of 5% Si, while, on the contrary, solution was slowed in the case of HCl. Aqueous solutions (up to 20%) of caustic alkalies and molten $KHSO_4$ do not decompose molybdenum disilicide, but molten caustic alkalies rapidly decompose it [23].

The resistance of molybdenum disilicide to molten metals is characterized by the following data [518]. Lead and tin do not react with it at 1000°, nor do sodium and bismuth [649]. Zinc does not decompose this disilicide, but slight solubility of the silicon (about 1%) in the metal is observed at 800°. Silver, gold, and mercury do not react with this compound. Aluminum decomposes it with the formation of molybdenum aluminide. Copper, iron, chromium, and platinum in the molten state react with molybdenum disilicide to form the silicides of these metals (partially binary).

According to Nowotny and Parthé (1954), metals of the first order form complex silicides with molybdenum disilicide. Metals of the second order react only when molybdenum forms chemical compounds with them. Molybdenum disilicide reduces Cr_2O_3 and, where little oxygen is available, SnO_2 also. With excess oxygen, metal oxides form silicates with the silicon of $MoSi_2$ and due to this, the resistance of this compound to metals is reduced.

Molybdenum disilicide does not react when fused with TiB_2, ZrB_2, and TiC, but reacts with Mo_2B, Mo_2C, and TaC under these conditions [660]. The composition of the reaction products has not yet been established. According to Kieffer, alloys of it with borides have a somewhat lower resistance to oxidation at high temperatures than pure molybdenum disilicide, though stable glassy films (boron compounds) are formed on them.

Refractory and current conducting articles may be made from molybdenum disilicide. Here it is possible to use various additives (for example, Al_2O_3) or binding metals (cobalt. nickel, etc.). Organic adhesives, for example, corn flour, are used to make the charge plastic [644]. Besides hot and normal pressing, ceramic casting may be used for making articles [652] and this makes it possible to improve their sintering. Oxide additives in molybdenum disilicide are only effective when they are not reduced by it. We should mention that in this respect, molybdenum disilicide reacts almost like elementary silicon. Al_2O_3 is usually used as an additive (it might be possible to use ZrO_2 and ThO_2 as additives). On the other hand, chromium oxide, Cr_2O_3, forms complex silicides and silica when heated with $MoSi_2$. At high temperatures, molybdenum disilicide forms SiO with silica. Oxides of Group IIA metals produce partial crystallization of the protective silica film on this silicide with a decrease in oxidation resistance at high temperatures, while Al_2O_3 has no negative effect.

Zirconium dioxide forms $ZrSiO_4$ in the film, which decomposes to the oxides again at 1500°. This process destroys the protective film.

The fracture strength of articles made from 75% $MoSi_2$ and 25% Al_2O_3 is 1547–2109 kg/cm^2. These articles have good heat and oxidation resistance up to 1480° and are considered better than articles of pure molybdenum disilicide or even titanium carbide with a cobalt binder [644]. The preparation of analogous articles has been proposed [650] from mixtures containing 15.5–77.7% $MoSi_2$, 21.5–81.5% Al_2O_3, and 0.8–3% CaO, which are pressed under 7750 kg/cm^2, dried, and preliminarily fired at 900–1350°, then machined, sintered at 1400 – 1600°, and additionally annealed at 1000–1500°. Such articles have a greater heat resistance, the higher their molybdenum disilicide content is. With a composition of 77.7% $MoSi_2$. 21.5% Al_2O_3, and 0.8% CaO, the fracture strength of articles was 246 kg/cm^2. The melting point was above 1800°. They were stable in air up to 1700°.

Metal additives in molybdenum disilicide articles have also been recommended, especially in the patent literature. Such articles of molybdenum disilicide with cobalt added had a bending strength of 4219 kg/cm^2 and with nickel added, 2714 kg/cm^2 [644]. Like iron, these metals must react with molybdenum disilicide to form cobalt and nickel silicides and also complex silicides. Silver, which does not react with molybdenum disilicide, is used as a binder for this silicide [656] to give articles with a high bending strength and good resistance to oxidation. Such articles may be used up to 900° [652].

The considerable oxidation resistance of molybdenum disilicide at high temperatures has promoted the development of methods for covering molybdenum with a protective layer of this silicide. Various methods have been tested: immersing molybdenum in silicon

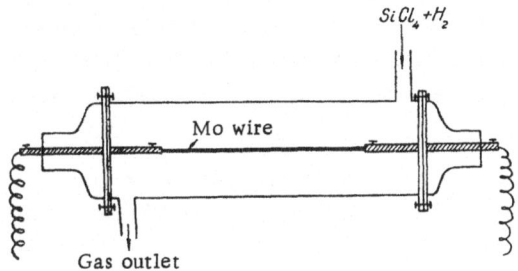

Fig. 92. Plan of apparatus for silicidation of molybdenum wire.

at high temperatures, atomization of silicon, and annealing molybdenum in a covering of powdered silicon. One of the most effective methods was found to be silicidation of molybdenum wire, bars, or tubes, heated to 1100–1800°, with a mixture of $SiCl_4$ vapor and hydrogen. The details of this method have been described [519, 520, 658]. A plan of the apparatus used is shown in Fig. 92.

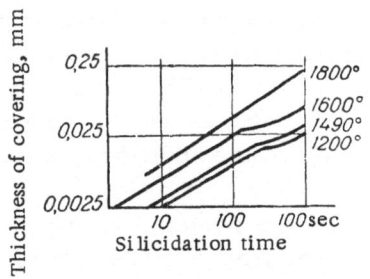

Fig. 93. Thickness of silicide layer formed on molybdenum by treatment with $SiCl_4$ vapor and H_2 in relation to the duration of the process at various temperatures, according to Beidler, Powell, Campbell, and Intema (1951).

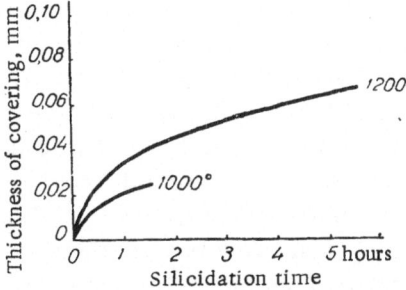

Fig. 94. Thickness of silicide layer formed on molybdenum by treatment with $SiCl_4$ vapor and H_2 in relation to the duration of the process at various temperatures, according to Fitzer (1952).

The molybdenum wire was heated with an electric current. The optimal temperature was about 1600° and the time, up to 100 sec. The relation between the thickness of the silicide coating on the molybdenum and time and temperature is presented in Figs. 93 and 94. From these figures it follows that the thickness of the covering increases more with a rise in temperature than with an increase in the silicidation time.

Fitzer [520] examined some thermodynamic problems in the silicidation of molybdenum.

A study of the operation time τ (hours) of molybdenum wire in relation to the thickness of the silicide covering b (mm) showed the relation could be expressed by the equation [519]

$$\tau = Kb^n,$$

where K and n are constants depending on the operation temperature. At 1200°, $K = 12,460$ and $n = 1.65$; at 1700°, $K = 2565$ and $n = 1.51$.

Silicidation of molybdenum makes it possible to obtain furnace heater elements whose operation time at 1700° with a silicide layer 0.125 mm thick is more than 1000 hours [519], but this requires further checking. This method may also be used for the production of parts for the manufacture of combustion chambers of motors, gas turbine blades, rocket nozzles, etc. [518, 519, 688]. Rubber and thermosetting materials are used for the dispersion of molybdenum disilicide [686], which is then deposited on graphite and this makes it possible to protect the latter from oxidation up to 1500°.

It is possible to make heaters from molybdenum disilicide both by hot-pressing and by ceramic methods, including molding [509, 517, 652, 657–659], for operation in air up to 1700°. The addition of stable oxides (of the Al_2O_3 and ThO_2 type) is used to decrease the electroconductivity of this disilicide and the addition of metals, to increase the strength [509].

When molybdenum disilicide heaters are cooled, the protective silica film formed on them cracks partially, but the cracks are removed during subsequent heating. Cracks are not formed in the protective film during cooling from 1700 to 900°. During the use of molybdenum-disilicide electric heaters, it is necessary to take special measures to prevent crumbling of a zone heated to 400–600°. Such additives as ZrO_2 and SiC considerably accelerate crumbling of articles made from this silicide under the conditions indicated. A preliminary heating of the articles above 1400° leads to the formation of a glassy film on them, which protects them from

oxidation at comparatively low temperatures. To decrease the crack-
ing of this film on cooling, it is recommended that shaped (e.g.,
U-shaped) heaters are made by ceramic casting. In this way, the
porosity of the parts may be reduced to 5-7% instead of the usual
20% [652]. At the present time molybdenum disilicide electric heaters are being manufactured, for example, by Fitzer (Austria) in
cooperation with the firm of "Siemens—Planck" in Augsburg (Fed-
eral German Republic). The heaters made operate for more than
3250 hours in air at 1700° [652]. Molybdenum disilicide is also used
for welding graphite at 2150° [687].

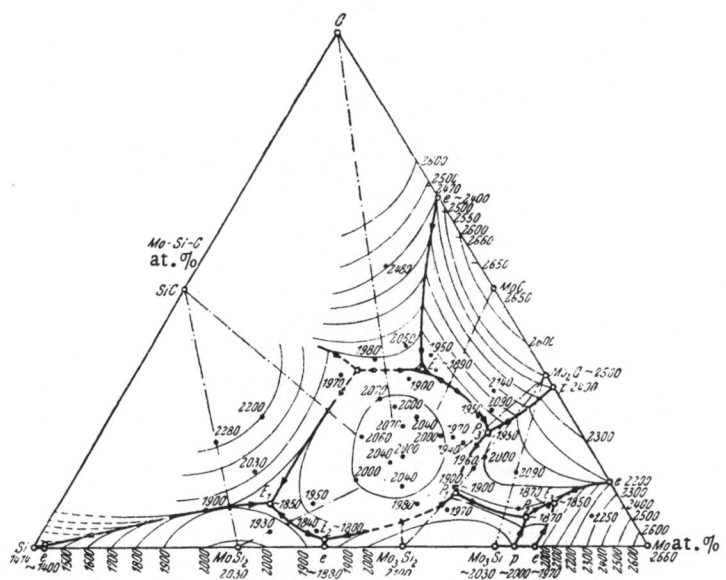

Fig. 95. Phase diagram of the Mo—Si—C system according to Nowotny, Parthé,
Kieffer, and Benesowsky (1954).

Thus, molybdenum silicides, especially $MoSi_2$, are promising
refractory and chemically stable materials for application in a
series of fields in new techniques. Therefore, their study and the
development of new methods of obtaining articles and coverings
from them are worthwhile.

The very valuable properties of molybdenum silicides have
also prompted the investigation of the Mo–Si–C system [153] and
the results of this are shown in Fig. 95. The ternary phase in
this system has the composition $Mo_6(Si_xMo_{1-x})(C_ySi_{1-y})_6$, where

$x = 0.10-0.55$, $y = 0.15-0.40$. This phase crystallizes according to the $D8_8$ type and has $a = 7.27$ kX, $c = 5.05$ kX for a high molybdenum content and $a = 7.27$ kX, $c = 4.99$ kX for the composition poorest in molybdenum. The specific gravity of this phase is 6.62, the microhardness 1460 kg/mm^2 and $\Delta H_{298\,\mathrm{form.}} = -21$ kcal/g-at. of Si. The Mo–SiC cross section of this system is shown in Fig. 96. It

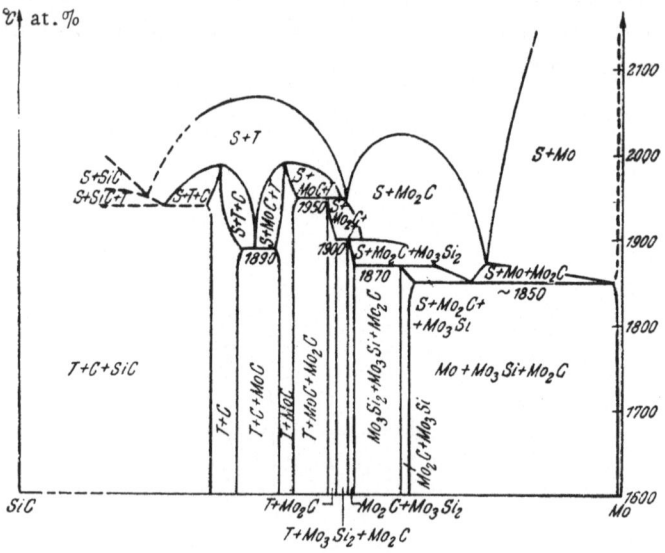

Fig. 96. Mo–SiC cross section of the phase diagram of the Mo–Si–C system.

is quite complex in structure. From this cross section and the diagram in Fig. 95 it follows that molybdenum cannot be a binding metal for carborundum since it does not give an equilibrium with the latter. Carborundum forms a simple binary subsystem with MoSi$_2$ with a eutectic melting at 1900°. In this connection, an investigation of combinations of carborundum and molybdenum disilicide is of interest in the technology of refractories and electric heating elements.

From the phase diagram of the Mo–Si–C system (Fig. 95) it follows that it is not expedient to sinter pure molybdenum disilicide in graphite crucibles due to the inevitable partial decomposition of this silicide and the formation of SiC and the ternary phase of this system gives a pseudobinary eutectic with carbon (melting point about 1950°) and is of interest as a binder in graphite articles.

The Mo–Si–B system, which has not been investigated yet, is also of interest. Separate experiments of Schwarzkopf and Glaser [27] showed that molybdenum disilicide reacts with boron at 1750° to form MoB. Systems including Mo–Si and another transition metal of Group IV, V, or VI are examined below.

W–Si SYSTEM. Tungsten silicides were first prepared by Moissan (1896) by the same methods as for molybdenum silicides. Later, tungsten silicide was also synthesized by Vigouroux (1898), who considered that this compound had the composition W_2Si_3.

Tungsten disilicide was prepared by Defacqz [521] by synthesis from silicon and tungsten in an electric furnace in the presence of copper silicides and also by aluminothermy.

The synthesis of tungsten silicides was also studied by Hönigschmid [486], Baroduc-Müller [444], and Frilley [496]. In 1913, Wedekind [27] obtained WSi_2 by hot-pressing at 1000–1200° and it was recently obtained by the same method by Cerwenka (1951). Brewer and his co-workers [487] synthesized two tungsten silicides with the compositions WSi_2 and $WSi_{0.70}$. Their eutectic points were as follows: $W–WSi_{0.70}–2020°$, $WSi_{0.70}–WSi_2–1890°$.

Hypotheses on the existence of tungsten silicides with the compositions W_2Si_3 and WSi_3 were not confirmed by the latest investigations.

The phase diagram of the W–Si system was worked out by Kieffer, Benesowsky, and Gallistl [522], who established the presence of two tungsten silicides, WSi_2 and W_3Si_2. Silicon dissolves in tungsten (up to 0.9%). The melting points of the eutectics in this system were found to be higher than determined by Brewer. According to the latest data [469, 470, 492], the tungsten-rich silicide should have the composition not of W_3Si_2, but W_5Si_3, isotypic with Mo_5Si_3. The presence of carbon, nitrogen, or oxygen stabilizes this phase.

If we adhere to the composition W_3Si_2, then the tetragonal lattice of this phase is characterized by the group D_{4h}^{18}, $a = 9.54$ kX, $c = 4.93$ kX, $z = 6$. Data on the lattice of the W_5Si_3 phase are presented in Table 2.

The structure of the WSi_2 phase, which is isotypic with $MoSi_2$ (see Table 2), was investigated by Zachariasen [508] and the heat of formation was determined by Robins and Jenkins [624].

Considering the similarity between molybdenum and tungsten compounds, one would expect the existence of a silicide with its composition W_3Si, but this has not yet been prepared.

A phase diagram of the W—Si system, which takes into account the observations given is presented in Fig. 97 and requires no further explanation. The regions of homogeneity of tungsten silicides have not been established precisely yet.

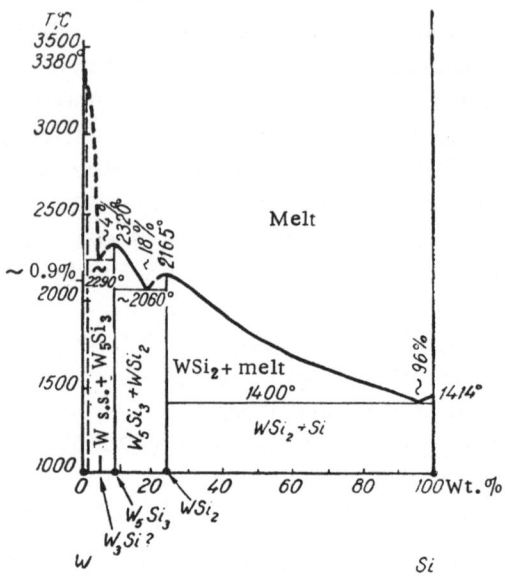

Fig. 97. Phase diagram of the W—Si system according to Kieffer, Benesowsky, and Gallistl (1952). The silicide "W_3Si_2" is replaced by W_5Si_3, according to the latest data. The silicide W_3Si is shown hypothetically.

Like molybdenum disilicide, but to a lesser extent, tungsten disilicide is resistant to oxidation during heating in air [522]; this is indicated by the change in weight of moldings of various disilicides, heated to 1200°:

Disilicide	Change in weight, $g/m^2 \cdot hour$	Disilicide	Change in weight, $g/m^2 \cdot hour$
$TiSi_2$	$+1$	$TaSi_2$	-128
$ZrSi_2$	$+105$	$CrSi_2$	$+128$
$ThSi_2$	$+182$	$MoSi_2$	$+1$
VSi_2	$+12$	WSi_2	-42
$NbSi_2$	-135		

The microhardness and electroconductivity of WSi_2 are less than those of $MoSi_2$ (see Table 2).

Tungsten disilicide reacts with fluorine at room temperature and with chlorine and bromine only on strong heating. It does not react noticeably with iodine. It ignites in oxygen when heated to red heat. It is very resistant to acids, but is decomposed slightly by concentrated hydrofluoric acid and dissolves readily in a mixture of HF and HNO_3. Caustic alkalies (10%) act slowly on this silicide. It decomposes readily in molten caustic alkalies, while molten $KHSO_4$ does not affect it, even at red heat.

The tungsten-rich silicide is also resistant to heating in air, but to a lesser extent than tungsten disilicide [26].

As with molybdenum, a protective silicide layer may be obtained on metallic tungsten, which makes it possible to use it as an electric heater up to 2000° [520].

Solid solutions of $MoSi_2$ and WSi_2 described below are very interesting in this respect.

These properties of tungsten disilicide are the reason for its value in high temperature techniques. This silicide has been investigated even less than $MoSi_2$.

The ternary system W–Si–C is analogous to the Mo–Si–C system [470]. It contains the ternary phase $W_{5-x} Si_{3-y} C_{x+y}$ of the $D8_8$ type with $a = 7.18$ kX, $c = 4.84$ kX.

Some Data on Ternary Systems Formed by Silicon with Transition Metals of Groups IV, V, and VI

In connection with the valuable technical properties of silicides of transition metals of Groups IV, V, and VI, preliminary investigations have also been made of some ternary systems of these elements with silicon or their cross sections. In most cases, the full phase diagrams of these systems have not been worked out yet.

Ti–Zr–Si SYSTEM. The metals of the Ti–Zr–Si system form a continuous series of solid solutions with a minimum on the liquidus curve [465]. The silicides Ti_5Si_3 and Zr_5Si_3 apparently give a continuous series of solid solutions. The $TiSi_2$–$ZrSi_2$ cross section is a pseudobinary system since $ZrSi_2$ melts incongruently (see Fig. 71). Here two solid solutions are formed [458, 525]: up to 5 mol.% $ZrSi_2$ in titanium disilicide and up to 55 mol.% $TiSi_2$ in zirconium disilicide. The system is heterogeneous with a 55–95 mol.% $TiSi_2$ content. The similarity of the melting points of $TiSi_2$ (1540°)

and $ZrSi_2$ (1520°) indicates that the melting points of their solid solutions cannot depend essentially on composition. The dimorphism of $TiSi_2$, which was not considered in investigations of the $TiSi_2$–$ZrSi_2$ cross section must complicate the interaction of these phases.

The oxidation resistance of $TiSi_2$–$ZrSi_2$ alloys at 1300° in a stream of air in relation to composition is characterized by the data [458] presented in Table 27. The table shows that with an increase in the $ZrSi_2$ content, the oxidation resistance of the $TiSi_2$–$ZrSi_2$ system is reduced. The friability of the alloys also increases at the same time [458].

Table 27

Oxidation Resistance of Alloys of the $TiSi_2$–$ZrSi_2$ and $ZrSi_2$–VSi_2 Systems at 1300° in a Stream of Air

Composition of alloys, mol. %			Change in weight, $g/m^2 \cdot hour$
$TiSi_2$	$ZrSi_2$	VSi_2	
100	—	—	+1
80	20	—	+2.5
60	40	—	+3
40	60	—	+6
20	80	—	−31
—	100	—	+48
—	80	20	−202.5
—	60	40	+30
—	40	60	+52
—	20	80	+16
—	—	100	+1

Ti–V–Si SYSTEM. Titanium and vanadium form a continuous series of solid solutions with a minimum on the liquidus curve [449]. Only the $TiSi_2$–VSi_2 cross section has been investigated in the ternary system Ti–V–Si [509, 525]. Titanium disilicide in the solid phase dissolves less than 5 mol. % of VSi_2 and vanadium disilicide, up to 85 mol. % of $TiSi_2$. It was established that there is a minimum (or a eutectic) on the liquidus curve of this system at 1400° and a $TiSi_2$ content of 80 mol. %. The decrease in the size of the two-phase region in the $TiSi_2$–VSi_2 system in comparison with the $TiSi_2$–$ZrSi_2$ system is caused by the great similarity of the atomic radii of titanium and vanadium in comparison with titanium and zirconium. Solid solutions of the (V, Ti) Si_2 type have a hexagonal structure. Most of the properties of $TiSi_2$–VSi_2 alloys change almost linearly with composition. Both the conductivity and the melting point have a minimum.

The Ti–Nb–Si system has not been studied at all.

Ti–Ta–Si SYSTEM. Titanium and tantalum form a continuous series of solid solutions [636]. The disilicides $TiSi_2$ and $TaSi_2$, however, form solid solutions only over the range of $TiSi_2$ contents of 0–50 mol.%. The phase diagram of the pseudobinary system $TaSi_2$–$TiSi_2$, according to Kudielka and Nowotny [635], is presented in Fig. 98. This figure shows that the presence of $TiSi_2$ considerably lowers the melting point of $TaSi_2$.

Ti–Cr–Si SYSTEM. Above 1350°, titanium and chromium give solid solutions and the chemical compound $TiCr_2$ is formed from them in the solid phase [449]. Only the disilicide section has been investigated in the Ti–Cr–Si system and it was established [499, 524] that at 1300° about 5 mol.% of $CrSi_2$ dissolves in $TiSi_2$, while on the $CrSi_2$ side the region of solid solutions extends approximately to 90 mol.% of $TiSi_2$. Thus at 1300° there is only a narrow two-phase region in this system (90–95 mol.% of $TiSi_2$). Investigation of the melting point in the $TiSi_2$–$CrSi_2$ sys-

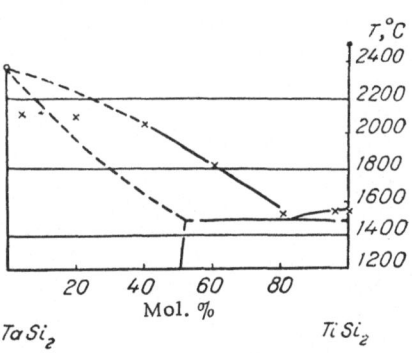

Fig. 98. Phase diagram of the $TaSi_2$–$TiSi_2$ system, according to Kudielka and Nowotny (1956).

tem showed [635] that there is a eutectic here at 95 mol.% of $TiSi_2$. The properties of chromium and titanium disilicide alloys change almost linearly with composition [509].

Ti–Mo–Si SYSTEM. Molybdenum and titanium form a continuous series of solid solutions [449]. A study of the Ti_5Si_3–Mo_5Si_3 system on samples pressed at 2000° showed [469] that despite the similarity of the atomic radii, titanium will only replace molybdenum up to 50 at.%. The carbon-containing hexagonal phase of Mo_5Si_3 apparently forms a continuous series of solid solutions with Ti_5Si_3. An ordered structure of the D_{6h}^4 type is obtained at the composition $Ti_3Mo_2Si_3$. Titanium disilicide hardly forms solid solutions in molybdenum disilicide at 1300–1500° [523]. At a $TiSi_2$ content of more than 40 mol.%, a new hexagonal phase appears ($CrSi_2$ type), which has a structure intermediate between $TiSi_2$ and $MoSi_2$. Apparently, a small amount of $MoSi_2$ (a few molecular percents) leads to the appearance of a new modification of $TiSi_2$. At more than 40 mol.% of $TiSi_2$ there is an almost linear change in the volume of the elementary cell of the solid solution in relation to composition. The size of the edge of the elementary cell of such solid solutions is as follows:

Composition	a	c
$(Mo_{0.6}Ti_{0.4})Si_2$	4.644	6.490
$(Mo_{0.2}Ti_{0.8})Si_2$	4.690	6.511

This phase of the $TiSi_2$–$MoSi_2$ system can also be considered as an intermetallide with a wide region of homogeneity and with properties changing with composition as for a solid solution.

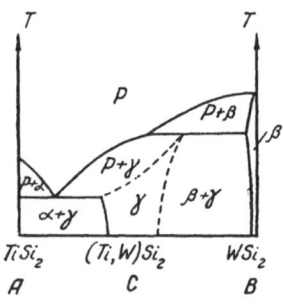

Fig. 99. Phase diagram of the $TiSi_2$–WSi_2 system, according to Kudielka and Nowotny (1956).

The melting points of the Ti–Mo–Si system were studied for the $TiSi_2$–$MoSi_2$ cross section during an investigation of fusibility in the $TiSi_2$–$MoSi_2$–WSi_2 system [523]. This cross section has a peritectic and a eutectic. The peritectic corresponds to fusion of the solid solution. The TiSi phase decomposes when heated with molybdenum, while Ta_5Si_3 is stable under these conditions. Titanium disilicide reacts with molybdenum to form Ti_5Si_3 [689].

Ti–W–Si SYSTEM. Titanium and tungsten form two series of solid solutions [527] with a peritectic at 1880°. In the Ti_5Si_3–W_5Si_3 cross section there is mixing [469] up to 25 mol.% of W_5Si_3. At a great tungsten silicide content, a tetragonal phase is present which was identified as W_3Si_2. Titanium disilicide does not give solid solutions in tungsten disilicide [523], but at a $TiSi_2$ content of more than 58 mol.%, a new phase of the $CrSi_2$ type is formed, as in the $TiSi_2$–$MoSi_2$ system, and the region of homogeneity of this extends almost up to pure $TiSi_2$. At the composition $W_{0.4}Ti_{0.6}Si_2$, the lattice constants are as follows: $a = 4.680$ kX, $c = 6.470$ kX. According to Kudielka and Nowotny [635], there are solid solutions here from 5 to 42 mol.% WSi_2 at 1300°.

The schematic structure of the $TiSi_2$–WSi_2 cross section, according to Kudielka and Nowotny [635], is presented in Fig. 99. The symbols on this figure are as follows: α) $TiSi_2$; γ) solid solution (Ti, W)Si_2; β) solid solution (W, Ti)Si_2; and P) melt. Thus, the γ-phase melts incongruently at about 1550° and the eutectic is below 1500°.

Zr–V–Si SYSTEM. Zirconium and vanadium form a eutectic system with the chemical compound ZrV_2 [465]. Carbon-stabilized V_5Si_3 and Zr_5Si_3 form solid solutions with limited solubility. At a Zr_5Si_3 content of 20–50 mol. %, there is a break in the solubility

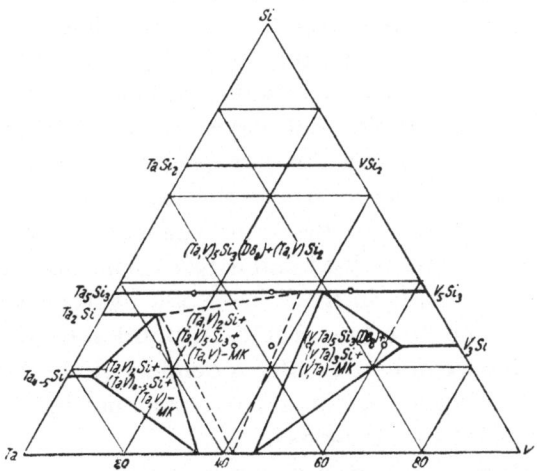

Fig. 100. Crystallization fields in the V—Ta—Si system, according to Nowotny, Lux, and Kudielka (1956).

[471]. Zirconium and vanadium disilicides hardly form solid solutions [458, 525, 635], which is caused by the great difference in the atomic radii of vanadium and zirconium.

The friability of $ZrSi_2$–VSi_2 alloys decreases with an increase in the VSi_2 content. The resistance of these alloys to oxidation is characterized by the data in Table 27.

The same table shows that VSi_2–$ZrSi_2$ alloys are no better than pure VSi_2 as regard resistance to oxidation.

The Zr–Nb–Si, Zr–Ta–Si, Zr–Cr–Si, Zr–Mo–Si, and Zr–W–Si systems have not been studied yet.

V–Nb–Si SYSTEM. Vanadium and niobium form a continuous series of solid solutions with a minimum on the liquidus curve at 35 wt. % Nb and 1810° [528].

The carbon-stabilized V_5Si_3 and Nb_5Si_3 phases form a continuous series of solid solutions as was established by x-ray determinations of the elementary cell constants [471]. Vanadium and niobium disilicides also form a continuous series of solid solutions without a change in the type of structure. The properties of solid solutions of vanadium and niobium silicides, including even their melting points, have not been studied yet.

V–Ta–Si SYSTEM. The V–Ta–Si system has been investigated very little as yet. Vanadium and tantalum disilicides form a continuous series of solid solutions, while Ta_5Si_3 and V_5Si_3 do so with a small break. A hypothetical arrangement of the separate phase fields in this system is presented in Fig. 100 [634].

V–Cr–Si SYSTEM. Only the V_3Si–Cr_3Si cross section of the V–Cr–Si system has been partially studied at 1300° [525]. These two silicides are isotypic and have similar cell constants (see Table 2). Therefore a continuous series of solid solutions is formed here, though their properties have not been studied yet.

V–Mo–Si SYSTEM. For the same reasons as for V_3Si and Cr_3Si (see above), the silicides V_3Si and Mo_3Si form a continuous series of solid solutions [525]. In the VSi_2–$MoSi_2$ system, which crystallizes as a different type (see Table 2), solid solutions are formed with a two-phase region at an $MoSi_2$ content of 65–97 mol.% [471]. The phase with a high VSi_2 content is hexagonal [640] and that with a high $MoSi_2$ content is tetragonal.

The properties of alloys of vanadium and molybdenum silicides have not been studied yet (including melting point), though they would be of interest, especially in the case of the $MoSi_2$–VSi_2 system.

The Nb–Ta–Si, Nb–Cr–Si, Nb–Mo–Si and Nb–W–Si systems have not been studied yet.

Ta–Cr–Si SYSTEM. The Ta–Cr system forms solid solutions and the chemical compound $TaCr_2$, which melts incongruently at 1940° [529]. Due to the large difference in the values of the lattice constants a (see Table 2), the isotypic silicides $TaSi_2$ and $CrSi_2$ form solid solutions at 1300° with a two-phase region at a $TaSi_2$ content of 12–60 mol.% [524, 635]. The properties of these alloys have not been studied yet. In the $TaSi_2$–$CrSi_2$ cross section there is a eutectic (25 mol.% $TaSi_2$), melting at 1400° [634].

Ta–Mo–Si and Ta–W–Si SYSTEMS. These systems have not been studied completely yet. Only their disilicide cross sections are known [635].

At 1300° tantalum disilicide forms solid solutions with $MoSi_2$ with a break in solubility at an $MoSi_2$ content of 56–84 mol. %. The eutectic in the $TaSi_2$–$MoSi_2$ system contains about 68 mol. % $MoSi_2$ and melts slightly above 2000°.

Tantalum and tungsten disilicides form solid solutions at 1300° with a break in solubility at a WSi_2 content of about 30–76 mol. %. The phase diagram of the pseudobinary system $TaSi_2$–WSi_2 is presented in Fig. 101, which shows that the eutectic here occurs at 40 mol. % WSi_2 and melts at 1900°.

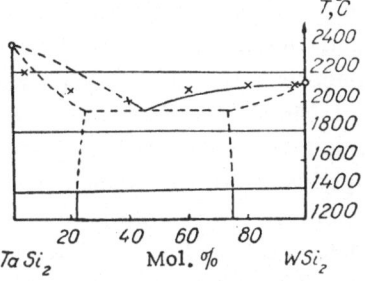

Fig. 101. Phase diagram of the $TaSi_2$–WSi_2 system, according to Kudielka and Nowotny (1956).

Thus, the addition of $TaSi_2$ to $MoSi_2$ or WSi_2 does not raise their melting points and, from this point of view, is of no interest in technology.

Cr–Mo–Si SYSTEM. Chromium forms solid solutions with molybdenum with a minimum in the liquidus curve at 25 wt. % Mo and a melting point of 1700° [530].

The isotypic phases with similar lattice constants, Cr_3Si and Mo_3Si, form a continuous series of solid solutions at 1300° [499, 524]. However, due to the volatility of MoO_3 and the poor sinterability of Cr_2O_3, formed during the oxidation of these solid solutions, the latter are unstable at high temperatures in air. For the reasons given, their weight changes little, though destruction proceeds rapidly [517]. Therefore, solid solutions of the Cr_3Si–Mo_3Si series are not promising refractory materials.

Chromium and molybdenum disilicides crystallize differently (see Table 2). At 1300° the $CrSi_2$–$MoSi_2$ system forms hexagonal solid solutions of (Cr, Mo) Si_2 and tetragonal ones of (Mo, Cr) Si_2 with a solubility break at a $CrSi_2$ content of 46–68 mol. %.

For the reasons given above, these solid solutions also are hardly of technical interest, the more so as $CrSi_2$ lowers the melting point of $MoSi_2$ [635].

Cr–W–Si SYSTEM. Chromium forms solid solutions with tungsten with a minimum on the liquidus curve at 8 at. % W and a melting point of about 1800°; these solid solutions decompose at temperatures below 1550° [530].

Only the $CrSi_2$–WSi_2 cross section has been studied in the Cr–W–Si system [523, 635]. Chromium and tungsten disilicides, which crystallize in different systems (see Table 2), form two types of solid solutions: hexagonal (Cr, W) Si_2 and tetragonal (W, Cr) Si_2. A solubility break at 1300° occurs at a WSi_2 content of 16–30 mol. %. The properties of these solid solutions have been studied very little, though they can hardly be of particular interest due to the ready oxidizability of chromium disilicide on heating and the fall in the melting point of tungsten disilicide in the presence of chromium disilicide [635].

Mo–W–Si SYSTEM. Molybdenum and tungsten form a continuous series of solid solutions [28]. The isotypic tungsten and molybdenum disilicides also form a continuous series of solid solutions at 1300° [523]. In connection with the very high resistance of both disilicides to oxidation in air when heated, this property of the disilicides and their solid solutions was studied [522] with the latter heated to 1500° in a stream of air (heating time 4 hours). Depending on the composition, different results were obtained for the change in weight and these are presented in Table 28.

Table 28

Change in Weight of (Mo, W) Si_2 Solid Solutions
When Heated in Air

Composition, %		Change in weight, g/cm^2
$MoSi_2$	WSi_2	
100	—	+0.0015
80	20	+0.0022
60	40	—0.0026
40	60	—0.014
20	80	—0.065
—	100	—0.023

From the table it follows that at a WSi_2 content of about 30%, the oxidation of (Mo, W) Si_2 solid solutions under these conditions should not be accompanied by changes in weight. Considering that the melting point is higher than that of pure $MoSi_2$ [523], such a solid solution is of interest for use, for example, as electric heater elements. The whole Mo–W–Si system undoubtedly deserves a thorough study.

SOME DISILICIDE CROSS SECTIONS OF QUATERNARY SYSTEMS. At the present time the following disilicide cross sections of quaternary systems have been studied [523, 635]: $TiSi_2-ZrSi_2-MoSi_2$, $TiSi_2-TaSi_2-CrSi_2$, $TiSi_2-TaSi_2-MoSi_2$, $TiSi_2-TaSi_2-WSi_2$, $TaSi_2-CrSi_2-MoSi_2$, $TiSi_2-CrSi_2-MoSi_2$, $TiSi_2-CrSi_2-WSi_2$, and $TiS_2-MoSi_2-WSi_2$. They are all characterized by the absence of

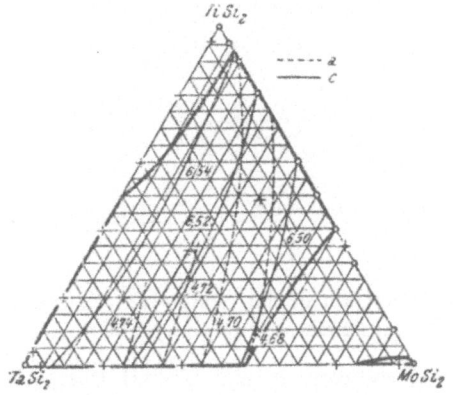

Fig. 102. Phase relations at 1300° in the $TiSi_2-TaSi_2-MoSi_2$ system, according to Kudielka and Nowotny (1956).

quaternary compounds and the presence of solid solution fields. Maxima were not observed on the melting point surfaces of any of the systems. An example of the structure of these disilicide cross sections is the pseudoternary system $TiSi_2-TaSi_2-MoSi_2$, whose phase fields at 1300° are shown in Fig. 102 and whose melting point diagram is given in Fig. 103 [635]. Figure 103 shows that from the point of view of melting point, the combinations of these disilicides are of no interest in the technology of refractories. The other properties of ternary alloys of disilicides of Group IV, V, and VI transition metals have hardly been studied yet.

TERNARY SYSTEMS FORMED BY TRANSITION METALS OF GROUPS IV, V, AND VI WITH SILICON AND BORON, CARBON, NITROGEN OR OXYGEN. With the exception of the more or less full investigation of the Mo–Si–C system described above and the partial investigation of the W–Si–C system, these systems still remain almost completely unstudied, despite their great interest. There are only the results of experiments [634] to determine the effect of boron, carbon, nitrogen, and oxygen in stabilizing the $D8_8$

phases of Me_5Si_3 and also to determine the existing phases in some Me–Si–C systems [689].

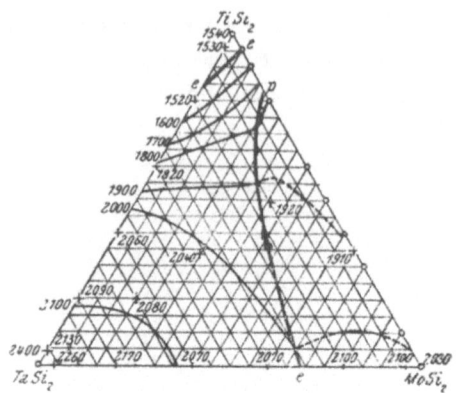

Fig. 103. Melting point diagram of the $TiSi_2$–$TaSi_2$–$MoSi_2$ system, according to Kudielka and Nowotny (1956).

In the Zr–Si–C system it was found that even at 2 at. % C the phases $ZrSi_2$, Zr_3Si, and Zr_6Si_5 disappear to leave only solid solutions of silicon in zirconium and carbon in Zr_5Si_3 (and a ternary phase of analogous composition) with a $D8_8$ type structure, $ZrSi$ and $ZrSi_2$, with which ZrC coexists.

Hence it follows, by the way, that carborundum must react with metallic zirconium on heating as with other transition metals of Groups IV, V, and VI.

During the absorption of carbon by Zr_5Si_3 there is an increase in the elementary cell constant c, while the constant a remains almost unchanged. Analogous phenomena occur in alloys of other metals also. There are no ternary compounds in the Zr–Si–C and Nb–Si–C systems. Zirconium disilicide coexists with ZrO_2, $ZrSiO_4$, and SiO_2 [689].

In the Ta–Si–C system the formation of a $D8_8$ phase may occur, for example, by the scheme

$$Ta_2Si + C \rightarrow TaC + Ta_5Si_3 (D8_8) + Si.$$

The ternary phase with the composition $Ta_{4.8}C_{0.5}Si_3$ has the following lattice constants: $a = 7.494$ kX, $c = 5.242$ kX, and a specific gravity of 12.48 [689].

At a C content of 10 at. %, chromium carbides were not formed in the Cr–Si–C system. The whole of the carbon was in the $D8_8$ phase.

A $D8_8$ phase is formed in the W–Si–C system with the greatest difficulty and then only at a high carbon content.

The following reactions occur in Me–Si–C systems:

$$Me + SiC \rightarrow \text{carbide } Me + Si,$$
$$Me + SiC \rightarrow Me_5Si_3 \ (C) + C.$$

At low temperatures, the second reaction is dominant for chromium and molybdenum, while for titanium, zirconium, tantalum, and tungsten, both these reactions proceed simultaneously. Carbides of transition metals are not stable at high temperatures relative to carbon-containing silicides. Tungsten carbide is relatively more stable than that of molybdenum and gives a $D8_8$ phase with difficulty.

In the Zr–Si–N system it was found that at an N content of 10 at. %, $D8_8$ phases (nitrogen-containing), ZrSi and ZrN were present.

When heated at 1400° with a mixture of Si_3N_4 and silicon, vanadium hydride changes predominantly into VN. Vanadium silicides are not formed. When treated with ammonia (1400°), the silicide V_5Si_3 ($T\,1$) gave predominantly VN and only a small amount of a phase with a $D8_8$ type structure.

Under the same conditions, the silicides Nb_5Si_3 ($T\,2$) and Ta_5Si_3 ($T\,2$) changed into a $D8_8$ phase and nitrides (NbN and TaN).

Nitrogen stabilizes the $D8_8$ phase much less than carbon for transition metals of Group VI.

The action of nitrogen on zirconium and tantalum silicides at relatively low temperatures leads to the formation of zirconium and tantalum nitrides. At very high temperatures, the silicides are more stable than the nitrides.

Oxygen is still less efficient at stabilizing $D8_8$ phases in Me–Si–O systems and sometimes these phases are not formed at all in its presence.

On the other hand, the addition of boron frequently stabilizes the $D8_8$ phase in Me–Si–B systems (especially in the Zr–Si–B system). The size of the elementary cell of boron-containing silicides of $D8_8$ phase increases.

The results of investigating the effect of carbon, boron, nitrogen, and oxygen on the stabilization of $D8_8$ silicide phases are presented in Table 29.

Thus, a series of polycomponent systems formed by transition metals of Groups IV, V, and VI with silicon are not only of scientific,

Table 29

Effect of Carbon, Boron, Nitrogen, and Oxygen (At. %)
on the Stabilization of $D8_8$ Silicide Phases [634]

Metal	C	B	N_2	O_2
Ti	+	Not investigated		
Zr	+(0,3)	+	+	+
Hf		Not investigated		
V	+(1)	+(5)	+	—
Nb	+(3)	+ (> 5)	+	—
Ta	+(5)	+ (> 5)	+(about 10)	—
Cr	+(5)	+ (> 5)	—	—
Mo	+(10)	—	—	—
W	+(more than 10)	—	—	—

Note.   + stabilizes, −does not stabilize.

but also of technical importance and this includes those with a high silicon content (up to the composition of disilicides). Together with this, polycomponent systems formed by transition metals of Groups IV, V, or VI with silicon, carbon, boron, and nitrogen are of exceptional interest, especially for refractory technology. The study of such systems has only just begun.

Systems Formed by Silicon with Transition Metals of Group VII

With the exception of the Mn–Si system, systems formed by silicon with transition metals of Group VII have hardly been studied at all. The structure of the electron shells of manganese, technetium, and rhenium is similar to that of transition metals of Group VI, though their atomic radii are different. Therefore, together with a certain similarity, there must be some difference in the structure of the phase diagrams with silicon and between the structures of silicides of Group VI and VII transition metals. All the manganese and rhenium silicides obtained have a metallic appearance.

Due to the lack of data, it is still impossible to make any general comments on the Mn–Si, Tc–Si, and Re–Si systems.

Mn–Si SYSTEM. In the middle of the last century Brunner (1858) synthesized manganese silicides from a mixture of CaF_2 and $MnCl_2$ by reduction with sodium in a Hessian crucible (reaction with the crucible). The product obtained ("Deville's manganese") as Wöhler showed, was manganese silicide. Manganese silicides were later obtained by different methods by Zefstre, Trust, Haute-feuille, and Vigouroux (1895) and others, but these silicides were not pure.

Vigouroux [531] considered that a silicide with the composition Mn_2Si was obtained by heating a mixture of silicon and manganese in an electric furnace, by reduction of a mixture of SiO_2 and manganese oxides with carbon or reduction of manganese oxides with silicon. Lebeau [478] also arrived at the same conclusion from his experiments. The monosilicide MnSi was isolated from iron alloys [532] and was also obtained synthetically by Lebeau. Manganese disilicide was obtained by de Chalmot [533] from a mixture of silicon, quartz, hausmannite, carbon, and lime in an electric furnace, but this product was contaminated with $FeSi_2$ and $CaSi_2$. Pure manganese disilicide was first obtained by Lebeau by fusing manganese with a large excess (ten-fold) of silicon and also by reduction of a mixture of manganese oxides and K_2SiF_6 with sodium.

Doerinckel [535] investigated the Mn–Si system by means of thermal analysis and microscopy. The starting materials were 98–99% pure. The alloys were prepared in a nitrogen atmosphere, which must have led to contamination of the preparations with nitrides, though this was not considered. Later, manganese silicides were synthesized by Frilley [496] and considerably later by Smith [539], but these authors gave little new information on the structure of the system.

Boren [497] investigated the Mn–Si system by x-ray methods, starting from very pure substances. He established the existence of the manganese silicides Mn_3Si, MnSi, and $MnSi_2$ and their structures (see Table 2). The solubility of manganese in silicon was very low.

Vogel and Bedarf [536] investigated the Mn–Si system by thermal analysis with a microscopic study of the alloys, which they prepared in an argon atmosphere but with only 97% pure manganese. It was found that Mn_3Si can dissolve up to 1% Mn. A silicide with the composition Mn_5Si_3 was found in the interval between Mn_3Si and MnSi. A thermal effect was observed at 980° and an Si content of 9–13%, though the reason for this was not established. Laves [537] investigated Vogel's preparations by x-ray methods. Westgren and Phragmen [538] studied the structure of manganese-rich alloys of the Mn–Si system even earlier.

These investigations established the phase diagram of the Mn–Si system presented in Fig. 104. The limits of homogeneity of the phases, especially Mn_5Si_3, MnSi, and $MnSi_2$, have not been determined accurately yet.

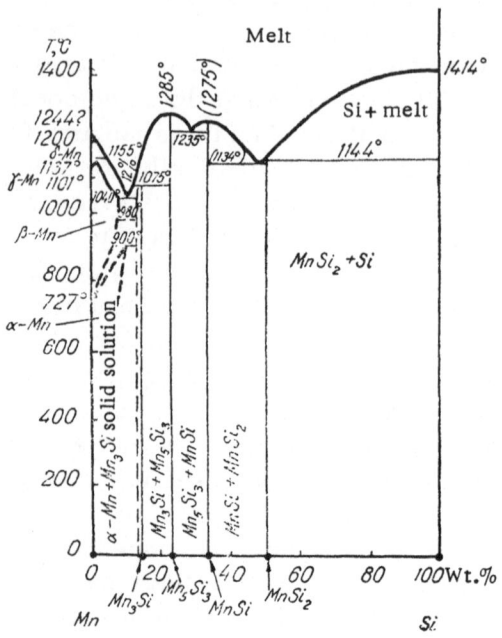

Fig. 104. Phase diagram of the Mn—Si system, according to Doerinckel (1906) and Vogel and Bedarf (1933—1934).

The silicide Mn_3Si is an electronic compound with a statistical distribution of manganese and silicon atoms in the lattice, whose structure is presented in Fig. 105 (α–Fe type). The manganese monosilicide MnSi is isotypic with CrSi (see above). Halogens readily decompose this silicide. Oxygen and water vapor begin to decompose it only at 1000°. Carbon reacts with molten manganese monosilicide to form SiC and Mn. HCl vapor also decomposes it readily. Nitric and sulfuric acids do not show any noticeable action on this silicide, even on heating, while concentrated hydrochloric acid decomposes it slowly at the surface only on prolonged boiling [23].

Manganese disilicide, $MnSi_2$, is decomposed by halogens and HCl with ignition when heated to red heat. Magnesium disilicide begins to be oxidized at 1000° in a stream of oxygen. Hydrofluoric acid readily dissolves it even in the cold, while concentrated hydrochloric and nitric acids show no noticeable action, even on boiling. Dilute solutions of caustic alkalies do not decompose this silicide, while concentrated ones react with it. Molten caustic alkalies decompose it very vigorously [23].

The chemical properties of the silicide Mn_5Si_3, which is isotypic with the previously described Me_5Si_3 silicides of Group IV, V, and VI transition metals, have not been studied yet.

Phragmen [637] found the silicide AlMnSi in the Al–Mn–Si system and this compound is not etched by acids, but is colored by an aqueous solution of NaOH (1 : 100). The lattice constants of this silicide were presented above (see Al–Si system).

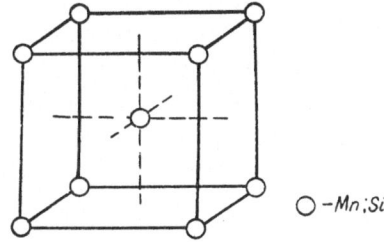

Fig. 105. Crystal lattice of Mn_3Si. Space group O_h^9—im3m; $a = 2.65$ kX, $z = 2$.

Manganese silicides have no special application. However, the Mn–Si system is of very great importance as a part of polycomponent systems, including those with iron (silicon spiegel). Many alloys based on them are made and these are widely used in technology and are described in detail in courses on metallography (see for example [79]).

Tc–Si SYSTEM. The Tc–Si system has not been studied at all yet. Technetium silicides have not been prepared yet. In analogy with the Mn–Si and Re–Si systems, it can be surmised that technetium silicides with the composition Tc_3Si (8.6% Si), TcSi (22.1% Si), $TcSi_2$ (36.2% Si) and perhaps Tc_5Si_3 (14.6% Si) should exist.

Due to the characteristics of technetium (complex preparation and radioactivity of the metal), technetium silicides will hardly have practical use.

Re–Si SYSTEM. The phase diagram of the very interesting Re–Si system has not been established yet. Rhenium disilicide, $ReSi_2$, was first obtained by Wallbaum [484], who established its structure, which was found to be isotypic with $MoSi_2$ (see Table 2). Searcy and McNees [540] used synthesis from the elements in graphite crucibles at 1600° in vacuum to prepare two new rhenium silicides, Re_3Si and ReSi, and measured their vapor pressure in relation to temperature. These investigators established that the regions of homogeneity of rhenium silicides have very narrow limits, which have not yet been determined accurately.

Due to the volatility of silicon at high temperatures, the following reactions occur:

$$\text{Re Si}_{2\text{sol.}} \rightarrow \text{Re Si}_{\text{sol.}} + \text{Si}_{\text{gas}},$$

$$\text{Re Si}_{\text{sol.}} \rightarrow \text{Re}_3 \text{Si}_{\text{sol.}} + \text{Si}_{\text{gas}},$$

$$\text{Re}_3 \text{Si}_{\text{sol.}} \rightarrow \text{Re}_{\text{sol.}} + \text{Si}_{\text{gas}}.$$

The transition of ReSi into Re_3Si was studied spectroscopically. The silicon vapor pressure was determined by Knudsen's method in graphite crucibles. The following values of the silicon vapor pressure, P, in relation to the absolute temperature, T, were obtained for the three given reactions:

$$\lg P = 7.518 - 25610 \cdot T^{-1},$$

$$\lg P = 7.444 - 25800 \cdot T^{-1},$$

$$\lg P = 5.953 - 24040 \cdot T^{-1}.$$

Considering the vapor pressure of silicon at its melting point, determined by Brewer (1950), the following relations were obtained between the free energy of formation of rhenium silicides, ΔF, and the absolute temperature, T, during formation from solid elements:

$$\text{Re}_3 \text{Si} \quad \Delta F = -21250 - 5.49 \cdot T,$$

$$\text{Re Si} \quad \Delta F = -26580 - 0.94 \cdot T,$$

$$\text{Re Si}_2 \quad \Delta F = -55030 + 0.74 \cdot T.$$

Re_3Si and ReSi are less stable thermodynamically than ReSi_2 at room temperature.

Searcy and McNees [540] conjectured that Re_3Si and ReSi should decompose with the formation of ReSi_2 and metallic rhenium. By means of x-ray powder patterns, these investigators [541] determined the crystal structure of ReSi, which was found to be cubic and isotypic with CrSi, MnSi, FeSi, and CoSi. The structure of Re_3Si has not been studied yet. Apparently it should be cubic like Mo_3Si and Mn_3Si.

The properties of rhenium silicides have not been investigated. It can be assumed that their melting points will be very high like those of molybdenum silicides. It is probable that rhenium silicides will also have other valuable properties for practical use. Therefore, it is very expedient to make a detailed study of the structure of the Re–Si system, including establishing its phase diagram and investigating the properties of rhenium silicides. It is also necessary to establish whether or not silicides with the composition Re_5Si_3 (8.3% Si) or Re_3Si_2 (9.2% Si) are formed in the Re–Si system similarly to the Mo–Si and Mn–Si systems. These investigations are equally interesting from the scientific and technical points of view.

Systems Formed by Silicon with Metals of Group VIII

Systems formed by silicon with metals of Group VIII can be conveniently divided into three classes according to their structure: 1) systems with metals of the iron subgroup, 2) light, and 3) heavy metals of the platinum group. Systems of the first class are very similar to the Mn–Si system and the last two differ sharply even in the composition of the silicides formed. Unfortunately the properties of silicides of platinum group metals have been studied very little as yet and therefore no generalizations are possible.

Silicides of iron group metals (iron, cobalt, and nickel), like the similar manganese silicides, have metallic properties and comparatively low melting points. Their chemical properties are similar. Platinum and palladium silicides, which have been studied in more detail, have a metallic appearance and are characterized by even lower melting points, indicating a weakening of the bonds in silicides of Group VIII metals in comparison with silicides of Group IV, V, and VI transition metals.

Systems Formed by Silicon with Metals of the Iron Subgroup

The atomic radii of iron (1.26 A), cobalt (1.26 A), and nickel (1.24 A) are almost identical. The structures of their electron shells differ only in the increasing filling of the M3d shell (six, seven, and eight electrons, respectively). In this respect, all three elements are very similar to manganese. Consequently, the structure of the phase diagram of the most fully investigated system, Fe–Si, is very similar to that of the Mn–Si system. The phase which was previously denoted by Fe_3Si_2 was found to have the composition Fe_5Si_3 and the structure of Mn_5Si_3 (see below). Other pairs of iron and manganese silicides have identical structures (see Table 2). The silicides Fe_3Si, Co_3Si, and Ni_3Si apparently should have the same structure. FeSi and CoSi have analogous structures, but not NiSi. The Ni_3Si_2 phase is hexagonal. Apparently, as with Fe_3Si_2, the composition Ni_5Si_3 should be ascribed to it (Si contents

24.19 and 22.31%, respectively), but this does not change the structure of the phase diagram of the Ni–Si system. In early investigations of the Co–Si system, a phase with the composition Co_3Si_2 was found, but this was then removed from the phase diagram by Vogel without grounds (see below). It was reported that the Co–Si system contains the silicide $CoSi_3$, which has no analogs in the Mn–Si, Fe–Si, and Ni–Si systems. We consider that both these peculiarities arise as a result of insufficient study of the Co–Si system. It is possible that the silicide $CoSi_3$ is only the limiting composition of the $CoSi_2$ phase, as Fe_2Si_5 is for $FeSi_2$ in the Fe–Si system. The silicide with the composition Co_5Si_3, like Fe_5Si_3 and Ni_3Si_2 (Ni_5Si_3), should be obtained only in the solid phase by prolonged annealing of alloys of appropriate composition, as was found originally. This silicide may simply have been missed in the thermal analysis of cobalt and silicon alloys.

The silicides Co_2Si and Ni_2Si have no analogs in the Mn–Si and Fe–Si systems. Their individuality should be rechecked allowing for the polymorphism of the Me_5Si_3 phases, which was described above for silicides of Group V and VI transition metals.

There is also no analog in other systems for the silicide Ni_5Si_2. Its individuality is doubtful and requires checking. ·

Cobalt and nickel disilicides do not have the same structure as iron disilicide. A polymorphic conversion of $FeSi_2$ was recently found. However, the structure of the modification thus formed has not been studied yet. Consequently it is impossible to compare this silicide with $CoSi_2$ and $NiSi_2$.

Despite the observations indicated, the lack of experimental data on the Co–Si and Ni–Si systems makes it impossible to make what are almost obviously necessary corrections in their phase diagrams. Consequently these systems are described below according to the old treatment, which should be regarded as only a first approximation, requiring refinement.

The thermodynamic characteristics of iron, cobalt, and nickel silicides have been studied very incompletely and also require refinement since, for example, a very improbable value (–160 kcal/mole) is given for the heat of formation of $FeSi_2$. The known thermodynamic constants of these silicides are presented in Table 2. The heat capacities of only a few silicides have been determined ($FeSi$, $NiSi$, and Ni_2Si); they have the following values:

$$\text{Fe Si} \quad C_p = 10.54 + 4.58 \cdot 10^{-3} \cdot T, \quad (273 - 903^\circ \text{ K});$$

$$\text{Ni Si} \quad C_p = 10.0 + 3.12 \cdot 10^{-3} \cdot T, \quad (273 - 1273^\circ \text{ K});$$

$$\text{Ni}_2 \text{Si} \quad C_p = 15.95 + 4.80 \cdot 10^{-3} \cdot T, \quad (473 - 900^\circ \text{K}).$$

The heats of formation of similar iron, cobalt, and nickel silicides (see Table 2) differ little.

There is only a weakly expressed tendency for a small increase in them from iron to nickel. The heats of formation of iron silicides, calculated for 1 g-at. of Si are almost identical, whereas for cobalt and nickel silicides they decrease noticeably as the silicon content increases. As the entropy values are not known, it is impossible to calculate the free energy of formation of these silicides and compare their thermodynamic stabilities. For FeSi, CoSi, and NiSi (the entropies have been calculated approximately in these cases) $\Delta F^\circ_{298^\circ}$ equals -22.8, -27.4, and -24.0 kcal/mole, respectively, i.e., very similar values.

Apparently, all the silicides of iron group metals are formed with a decrease in volume (see Table 2).

These general comments should be borne in mind in examining the systems described below.

Fe-Si SYSTEM. Alloys of iron and silicon (ferrosilicon) were first prepared in 1808 by coreduction of silica and iron oxides with carbon. St. Claire Deville and Caron and also Winkler worked in this field. The smelting of ferrosilicon began in Russia at the end of the Nineteenth Century in the Ural factory of "Porogi," close to the town of Satka [544]. Individual iron silicides were also known a long time ago. Hahn (1864) considered that he obtained a silicide with the composition Fe_2Si by the action of silicon and sodium on $FeCl_3$ and NaCl (with heating). Moissan (1895–1896) also considered that a silicide of this composition was formed by fusion of the components. Lebeau (1900–1902) and Vigouroux (1905) arrived at the same conclusion. However, no iron silicide of this composition was found in later investigations of the Fe-Si system by modern methods.

By reduction of a mixture of iron and quartz sand with carbon in an electric furnace, de Chalmot [548] obtained a silicide with the composition Fe_3Si_2. Until recently, this silicide was shown on phase diagrams of the Fe-Si system, but it was found to have the composition Fe_5Si_3 by more accurate investigations [549].

Iron monosilicide, FeSi, was first obtained by Hahn (1864) by reduction of iron and sodium chlorides with metallic sodium in the presence of Na_2SiF_6. This compound was also obtained by Lebeau (1901), who carried out many experiments on the Fe–Si system, and also Vanzetti (1906). Pure iron disilicide, $FeSi_2$, was first obtained by Lebeau [550] by direct synthesis from the elements in an electric furnace. This was the reason why Kurnakov and Urazov [566] named the corresponding ζ-phase in the Fe–Si system "lebeauite."

The later history of the synthesis and study of iron silicides is not of particular interest, the more so as it has already been

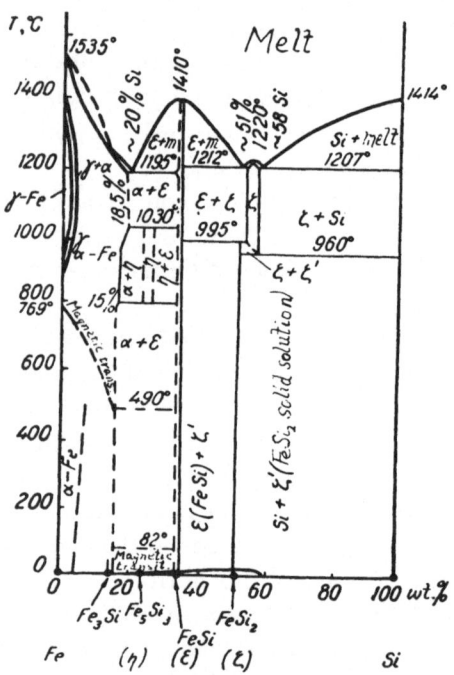

Fig. 106. Phase diagram of the Fe–Si system, according to Haughton, Becker (1930), and others, with corrections by Weill (1945), Smithells (1949), and Abrikosov (1956).

described in detail more than once [28, 551–554]. Therefore we can proceed directly to a discussion of modern data on the structure of the phase diagram of the Fe–Si system and its separate phases.

The Fe–FeSi section of the Fe–Si system has been investigated by many scientists, including Kurnakov and Urazov [555] and also Haughton and Becker [556]. The latter authors worked with the

purest starting materials. Therefore their data were used predominantly for the Fe–FeSi region of the phase diagram of the
Fe–Si system (Fig. 106). In this system, the γ-Fe phase has a
closed region of existence [556, 557, 559] and extends approximately
up to 2.5% Si at 1160° [558]. The change in the temperature of the
magnetic transition is according to the data of Haughton and Becker
[556].

The transition in the α-Fe solid solution field, demonstrated
by many investigators [553, 554, 557, 560], occurs as a result of
the formation of an Fe_3Si superlattice. This is accompanied by an
increase in electroconductivity and friability of the alloys. The
structure of this phase was determined by Phragmen [561] and
Jette [560]; the characteristics of the elementary cell are given
in Table 2. The region of existence of the Fe_3Si phase is not shown
on the phase diagram of the Fe–Si system as it has not been established adequately as yet [562, 563]. Although the new $α_1$- and $α_2$-
phases, which exist at temperatures above 1000°, have been shown,
this region was not determined accurately even in the re-examination of the phase diagram of the Fe–Si system in the region
containing 0–30% Si [549]. The $η$ phase, which was previously considered as Fe_3Si_2 (25.11% Si), was found to have the composition
Fe_5Si_3 (23.18% Si) with a structure of the Mn_5Si_3 type [549]. This
phase has the small stability region shown in Fig. 106. Its lattice
constants are presented in Table 2.

For the reaction in the liquid state

$$Fe + Si = FeSi$$

at working temperatures $ΔH° = -29$ kcal, the change in free energy

$$ΔF° = -28\ 500 - 0.64\ T.$$

The high negative values of $ΔF°$ show that iron monosilicide,
FeSi, is almost undissociated in the melt.

As Fig. 106 shows, the region of homogeneity of the δ-phase,
FeSi, is narrow. The structure of this phase (see Table 2) was
investigated by Phragmen [557], Boren [497], and also Wever and
Müller [564]. The region of FeSi–Si compositions of the Fe–Si
system includes only one intermediate ζ-phase, whose composition
is sometimes represented by the formula Fe_2Si_5 with an Si content
of 55.71% [556]. However, according to the x-ray investigations of
Phragmen [557], its composition corresponds to $FeSi_2$ with a de-
ficiency of iron atoms, which are replaced by silicon atoms. The

region of homogeneity of this phase has been studied by many investigators. Ageev and his co-workers [565] established that the region of homogeneity of lebeauite (ζ-phase) is at 53.0–57.7% Si. Ageev also demonstrated the presence of a transition in the solid phase above 900° with the formation of the new x-phase. By measuring the heat capacity of lebeauite, Serebrennikov and Gel'd [567] found that there are polymorphic conversions according to the scheme

$$\zeta_\gamma \overset{650°}{\underset{}{\rightleftarrows}} \zeta_\beta \overset{908°}{\underset{}{\rightleftarrows}} \zeta_\alpha .$$

The first conversion proceeds comparatively slowly in the reverse direction. The heat effect of the second conversion at 908° equals approximately 6–7 cal/g.

This conversion of lebeauite was investigated in more detail by Abrikosov [568], who found that the high-temperature conversion occurs here at 960°. He did not study the second conversion or the structure of the phases formed. Corrections have been introduced into the diagram in Fig. 106 according to the results of Abrikosov's investigations. The temperature dependence of the thermal expansion coefficient of lebeauite is as follows [50]:

$$\alpha\text{-lebeauite } \alpha \cdot 10^6 = 8.7175 + 6.125 \cdot 10^{-3} t,$$
$$\beta\text{-lebeauite } \alpha \cdot 10^6 = 7.7520 + 5.744 \cdot 10^{-3} t - 0.656 \ 10^{-6} t^2.$$

It should be noted that the low-temperature ζ'-phase is very close in composition to $FeSi_2$. Considering that $CoSi_2$ and $NiSi_2$ have structures of the CaF_2 type, it would be worthwhile establishing whether or not $FeSi_2$ has such a modification.

The solubility of iron in silicon is very low. By microscopic investigations, Murukami [569] established that it is not more than 4%. The x-ray investigations of Phragmen [557] showed that iron is practically insoluble in silicon. According to Abrikosov [568], even an alloy containing 99.5% Si contains a second phase.

Due to their great technical importance, the properties of Fe–Si alloys have been studied in great detail. The strength constants were studied by Smith and Daniels [570]. Iron silicides, especially $FeSi_2$, are very friable. Glaser and Iwanick [683] studied the process of the ordering of Fe–Si alloys up to the composition Fe_3Si.

The chemical properties of Fe_3Si have been studied very little. The silicide Fe_5Si_3 (Fe_3Si_2) is hard and friable, nonmagnetic and conducts an electric current well. The substance dissolves slowly in a mixture of HCl and HNO_3 and rapidly in HF. This silicide is readily decomposed by a molten mixture of Na_2CO_3 and $NaNO_3$ [23].

Iron monosilicide, FeSi, is decomposed by fluorine at room temperature and by chlorine and bromine with ignition at red heat. It is not oxidized by oxygen when heated to high temperatures. Nitric and sulfuric acids do not react with it at all. Hydrochloric acid dissolves it slowly, but completely. A mixture of HCl and HNO_3 reacts with iron monosilicide more strongly the more HCl it contains, but less than pure hydrochloric acid alone. Hydrofluoric acid dissolves it rapidly. Molten alkalies and also mixtures of alkali carbonates and nitrates readily decompose FeSi. This silicide also decomposes when fused with copper [23]. In a melt at 1500–1700°, the reaction corresponding to the following equation occurs [684]:

$$FeSi + 2FeO \rightleftarrows 3Fe + SiO_2.$$

Iron disilicide, $FeSi_2$, is dark gray in color and has a strong luster. In a finely powdered state it is decomposed by fluorine in the cold and by chlorine and bromine at red heat. This compound is oxidized on the surface in oxygen at 1200°. Apart from HF, none of the mineral acids (even concentrated ones and with heating) react with it. Hydrofluoric acid reacts with iron disilicide even in the cold and the latter dissolves rapidly on heating. Solutions of alkalies show no noticeable action on this silicide in the cold, but slowly decompose it on heating. Molten caustic alkalies decompose iron disilicide very rapidly [23]. When it is heated in air or oxygen in a mixture with oxides of rare earth metals, a reaction occurs at 400° and with BaO, at 329° with an explosion. Silicates are formed as a result [571].

Alloys of iron with silicon have very great technical importance. An Si content of 1–2% improves the strength characteristics of steel (spring steels 60C2, EI142, etc.). Transformer steel, which has a low residual magnetism, contains 4–4.5% Si. Alloys with an Si content of 14–18% have a high corrosion resistance.

The greatest use of iron silicides is the production of ferrosilicon, which is widely used in metallurgy for the deoxidation of many grades of steel and the preparation of low-carbon iron alloys. The over-all consumption of ferrosilicon is about 0.3% of steel smelted [544]. The composition of ferrosilicon is controlled in the factory by determination of the specific gravity of the alloy [544] and this indicates the silicon content approximately (Fig. 107).

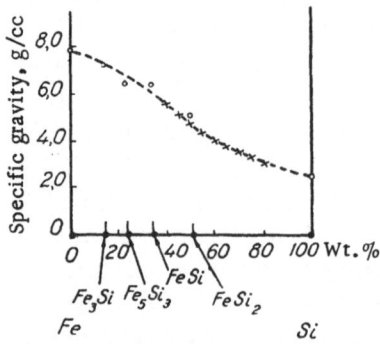

Fig. 107. Specific gravity of alloys of the Fe—Si system, according to Elyutin, Pavlov, and Levin (1951). The circles mark the specific gravity of silicides and pure components, calculated from the size of the elementary cell.

Ferrosilicon containing 9–17% Si is smelted in blast furnaces. Ferrosilicon richer in silicon is smelted from quartzite and iron turnings in electric shaft furnaces with carbon linings. Coke is used as the reducing agent. According to Elyutin [544], the cost of the electric power in the smelting of ferrosilicon is about 35–40% of the total expense. The specific consumption of electric power increases with an increase in the silicon content of the ferrosilicon (Table 30).

The technical and economic factors in the production of ferrosilicon are characterized by the data presented in Table 30.

The thermodynamics of the processes in ferrosilicon production were worked out by Elyutin [544] and Dashevskii [572]. The reduction of SiO_2 by carbon is facilitated by the presence of iron

Table 30

Technical and Economic Factors in the Production of One Ton of Ferrosilicon in Electrical Furnaces [544]

Material	Silicon content of alloy			
	20%	45%	75%	100%
Quartzite, kg	460	1050	1800	3000
Iron turnings, kg	900	600	260	—
Coke, kg	275	580	1000	—
Pitch coke, kg	—	—	—	500
Wood charcoal, kg	—	—	—	1300
Rammed electrodes, kg	12	30	50	—
Graphited electrodes, kg	—	—	—	120
Electric power, kw-hr	2100	4700	8500	14000

due to the formation of FeSi. Thermodynamic calculations [544] showed that the temperature at which the reduction of SiO_2 by carbon begins depends on the silicon content of ferrosilicon: this is shown by the following:

Silicon content of ferrosilicon, %	Temperature at which reduction of SiO_2 begins, °C
100	1541
75	1533
45	1501

In the preparation of ferrosilicon a considerable amount of silicon is volatilized, apparently in the form of SiO.

According to GOST 805-41 (blast furnace) and 1415-49 (electrothermal), ferrosilicon used in metallurgy must satisfy the requirements presented in Table 31.

Table 31

Chemical Composition Requirements for Ferrosilicon

Method of preparing ferrosilicon	Grade	Chemical composition, %				
		Si	Mn	Cr, not more than	P, not more than	S, not more than
Blast furnace.	FS_1	Not less than 13,10	3.00	—	0,15	0,04
» »	FS_2	9—13	3,00	—	0,15	0,04
Electrothermal	Si 90	87—95	0,5	0,2	0,04	0,04
»	Si 75	72—78	0,7	0,5	0,05	0,04
»	Si 45	43—50	0,8	0,5	0,05	0,04

Ferrosilicon conducts an electric current well. Its maximum hardness occurs at an Si content of 33%. Ferrosilicon is resistant to the action of various chemical reagents. Ferrosilicon up to an Si content of 50% is soluble in hydrochloric acid. Nitric and sulfuric acids hardly react with it, while HF dissolves it. The disintegration of ferrosilicon, accompanied by explosions, fires, and toxic effects, which are observed particularly during its sea transport, provoked an investigation of the reasons for the disintegration [566, 573]. It was established that completely pure ferrosilicon does not disintegrate. The disintegration phenomenon only occurs in moist air when the material contains about 3% Al, 0.04% P, the same amount of Ca, and 33–75% Si, especially 50–60% Si.

Consequently, ferrosilicon with an Si content of 50–70% is not produced. Rapid cooling prevents and slow cooling promotes disintegration of ferrosilicon.   The toxic effect occurs as a result of the formation of phosphine and AsH_3. At an Al content of more than 1.7%, a solid solution with an ε-phase structure is formed [573]. This solid solution is unstable and at an Si content of 50–60%, a ς-phase is formed from it in the solid state and this is accompanied by an increase in volume of almost 20.5%, which is the reason for disintegration of such ferrosilicon like unstable dicalcium silicate. In connection with this, the Fe–Al–Si system was studied [577]. During normal casting of 75% ferrosilicon, considerable liquation occurs and a zone is formed with a dangerous silicon content. To prevent this, pouring ferrosilicon into small metal molds with a massive, strongly cooled bottom plate [574] and making up the charge with components as free as possible from aluminum, phosphorus, and calcium [575] have been proposed. The high corrosion resistance of Fe–Si alloys is explained by the formation of an SiO_2 film on them [694]. Thermal diffusion silicidation of steel has been recommended [695].

As mentioned above, ferrosilicon is widely used in metallurgy as a reducer and in the smelting of silicomanganese, ferrotungsten, and ferromolybdenum. It has been proposed to use it as electrodes in the electrolysis of aqueous solutions as such electrodes are more resistant to attack than carbon ones. It has also been recommended for the manufacture of parts of chemical apparatus, for example, for the production of concentrated sulfuric acid. The negative property of ferrosilicon here is its very great friability. Statues and architectural features can be manufactured from 25% ferrosilicon since they have a beautiful external appearance after polishing. Silicon-rich ferrosilicon can be used for the preparation of $SiCl_4$ (see the preparation of silicon halides) and also hydrogen by the reaction

$$Si + 2NaOH + H_2O = Na_2SiO_3 + 2H_2.$$

The firm "Magnesital" [576] proposes to use the addition of 1–10% ferrosilicon, containing about 90% Si, to the mix for magnesite (periclase) refractories to obtain volume-constant (with a change in temperature) articles.

Many Me–Fe–Si systems are of great technical interest and the results of their investigation are used in industry. The following alloys serve as examples.

Silicomanganese, containing 14-20% Si, more than 60-65% Mn, and about 15-20% Fe (GOST 4756-49), is obtained by fusing ferrosilicon and ferromanganese in electric furnaces or by reduction of iron and manganese oxides and SiO_2. According to the given chemical composition, this alloy in the equilibrium state must contain two solid solutions: $(Mn, Fe)_3Si$ and $(Mn, Fe)_5Si_3$. Silicomanganese is used for the deoxidation of some grades of steel.

Phragmen [637] investigated the phase encountered in the Al-Fe-Si system and found here the silicide AlFeSi, which exists in three forms: cubic, monoclinic, and tetragonal (see above section on the Al-Si system). This system is primarily of interest in the production of alloys rich in aluminum and also in the manufacture of steel.

The Cr-Fe-Si system was investigated by Kurnakov [578] and Elyutin [544]. Alloys of the Cr-Fe-Si system with a low carbon content consist [544] of $CrSi_2$, $FeSi_2$, SiC, and Si. Ferrosilicochromium is of technical importance. In the slag process for obtaining it, a mixture of chromite and quartzite is reduced with carbon in an electric arc furnace. Chromium carbides are first formed and these then react with silicon to give silicides. The carbon separates in the free state and as partially bound SiC. In the slagless method of producing this alloy, the charge consists of quartzite, ferrochromium, and coke. The silicon content of ferrosilicochromium is usually 27-30%, but it also may reach 45-50% [544]. Ferrosilicochromium is used for the production of low-carbon ferrochromium.

For other systems with Fe-Si see [544, 579, 580, 637].

Co-Si SYSTEM. At the end of the last century Vigouroux (1895) obtained the silicide Co_2Si by heating silicon and cobalt in an electric furnace. Lebeau (1901) synthesized the monosilicide CoSi by the action of copper silicides on cobalt in an arc furnace. Lebeau (1902) was also the first to obtain cobalt disilicide, which he did from a mixture of copper silicide, cobalt, and silicon by heating it in a carbon crucible in an electric furnace. The first systematic investigation of the phase diagram of the Co-Si system was undertaken by Lewkonja [581], who used cobalt of 99.38% and silicon of 98.5% (about 1% Fe) purity. The alloys were heated first in hydrogen and then in a stream of nitrogen. Lewkonja investigated a total of 33 alloys, which covered all concentrations. The alloys were investigated by thermal analysis, microscopy, and magnetometry. The following intermediate phases were found: Co_2Si (straightforward maximum at 1327°); Co_3Si_2, formed as prismatic crystals in the solid phase at temperatures below 1215° from Co_2Si and $CoSi$; $CoSi$ (straightforward maximum at 1935°); $CoSi_2$ (incongruent melting at 1277°), and $CoSi_3$ (straightforward maximum at 1308°). In addition it was found that up to 7.5% of Si dissolves in cobalt and as a result the temperature of the magnetic transition is lowered. According to Lewkonja, up to 9% of cobalt dissolves in silicon. The modification changes of cobalt and the effect of silicon on them were not studied in these investigations.

The formation of the silicide Co_3Si_2 due to reaction in the solid phase at 1215° was also reported by Baroduc-Müller [444].

Boren [497] investigated alloys of cobalt and silicon by x-ray methods and found only two silicides: Co_2Si and $CoSi$; he established that Co_2Si has a rhombic structure and $CoSi$ is isotypic with $FeSi$ (see Table 2). This investigator showed that silicon is soluble in the hexagonal and cubic modifications of cobalt and also confirmed the formation of solid solutions of cobalt in silicon.

In connection with a study of the Fe–Co–Si system, Vogel and Rosenthal [583] reinvestigated the Co–CoSi section of the Co–Si system by thermal analysis and microscopy. The data they obtained differed from the results of the experiments of Lewkonja with respect to melting points (insignificant) and the existence of separate phases. These investigators found a new silicide, Co_3Si, formed peritectically at 1210° and decomposing at 1160° in the solid phase to a solid solution of cobalt and Co_2Si. It was found that Co_2Si exists in the form of two modifications with a transition point at 1320° for the pure substance and 1208° for a solid solution with silicon. The high-temperature modification (β-Co_2Si) contains up to 20.8% Si and the low-temperature α-form, only up to 19.8% Si. In the change of β-Co_2Si in a solid solution into the α-form, the excess silicon is liberated as $CoSi$. Vogel and Rosenthal investigated a total of just over 20 alloys. A study of the microstructure of an alloy with the composition Co_3Si_2, which had been cooled slowly but not annealed till equilibrium, showed that it consisted of two phases. On this basis alone, which is quite inadequate in our opinion, Vogel and Rosenthal concluded that Lewkonja's deductions on the existence of the silicide Co_3Si_2 were incorrect. Considering the structure of the similar systems Mn–Si, Fe–Si, and Ni–Si, it can be maintained that the phase Co_3Si_2, or, more accurately, Co_5Si_3, must exist in the Co–Si system, forming in the solid state like the analogous phases in the Fe–Si and Ni–Si systems.

The presence of the silicide $CoSi_3$, which was found by Lewkonja, but investigated very little by him, requires checking.

Considering all available data, Fig. 108 shows the phase diagram of the Co–Si system, which should be considered insufficiently

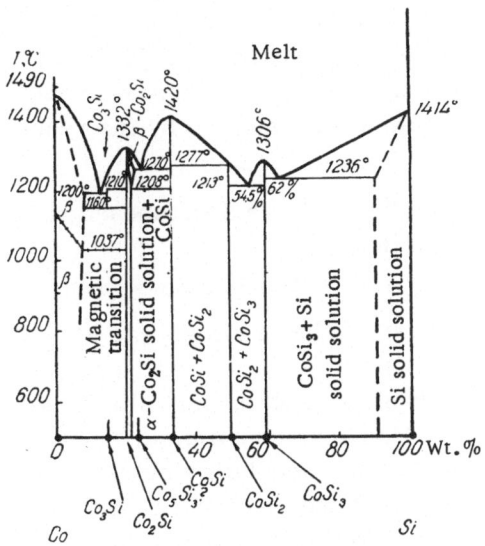

Fig. 108. Phase diagram of the Co–Si system according to Lewkonja (1908) and Vogel and Rosenthal (1934).

studied as yet. It requires careful investigation by modern methods. Apart from what has been said, the nature of the Co_2Si phase and whether or not it is actually Co_5Si_3 should be determined. It is also necessary to determine the limit of solubility of cobalt in silicon more accurately.

The crystal structure of Co_3Si has not been studied yet. This silicide should apparently be isotypic with Fe_3Si. The structure of Co_2Si was investigated by Boren [584] (the lattice constants found are presented in Table 2) and also by Geller and Wolontis [585]. The results of the two investigations agree in general. The silicide Co_2Si (low-temperature modification) is isomorphous with Rh_2Ge and somewhat reminiscent of FeB. The silicon atoms in it form zigzag chains. We consider that the data obtained on the structure of Co_2Si should be reexamined to determine the relation of this silicide to the possible Co_5Si_3 (Co_3Si_2).

The structure of CoSi has been studied more than once [497, 586] and was found to be isotopic with that of FeSi (see Table 2). Cobalt disilicide, $CoSi_2$, crystallizes according to the CaF_2 type [587].

The heats of formation of cobalt silicides were determined by Körber [588] and Oelsen [589].

Greenhouse and his co-workers [590] found that cobalt monosilicide, CoSi, is formed in the solid phase from silicon and cobalt at temperatures below 1510° and also by the reaction

$$SiC + Co = CoSi + C.$$

Titanium disilicide, $TiSi_2$, is decomposed at high temperatures by cobalt by the reaction

$$TiSi_2 + 2Co = 2CoSi + Ti.$$

The microhardness of Co–Si alloys was studied by Savitskii and Baron [633].

The chemical properties of cobalt silicides have been studied very little as yet. Cobalt monosilicide, CoSi, is decomposed by fluorine on gentle heating and by chlorine at dull red heat. This compound is slowly oxidized at 1200° in a stream of oxygen. Water vapor oxidizes it at 1200°, H_2S forms cobalt and silicon sulfides with it, and at red heat HF and HCl form cobalt and silicon fluorides and chlorides with cobalt monosilicide. CoSi changes into Co_2Si with the liberation of free silicon in the presence of molten silver. Cobalt monosilicide is resistant to the action of HNO_3 and concentrated H_2SO_4, but dissolves slowly in a mixture of HCl and HNO_3 and more rapidly in hydrochloric acid. Dilute solutions of caustic alkalies show no action on it, but concentrated solutions and molten alkalies decompose this silicide. Alkali carbonates act similarly. Potassium nitrate does not decompose cobalt monosilicide [23].

Fluorine reacts with cobalt disilicide, $CoSi_2$, on gentle heating and chlorine at 300°. Cobalt disilicide is oxidized on the surface at 1200° in oxygen. It is quite resistant to the action of sulfuric and nitric acids but decomposes slowly when boiled with concentrated hydrochloric acid. Hydrofluoric acid dissolves it rapidly. Dilute solutions of caustic alkalies show no action on cobalt disilicide, but concentrated solutions or molten caustic alkalies decompose this silicide [23].

Me–Co–Si systems are of interest in connection with the manufacture of refractory silicide cermets in which cobalt or its silicides could serve as a binder. However, these systems have been studied very little as yet.

Ni–Si SYSTEM. Nickel silicide, Ni_2Si, was obtained by Vigouroux (1895) from nickel and silicon at the temperature of a reverberatory furnace and in an electric furnace. Later, this silicide and also NiSi and Ni_3Si_2 were obtained by Frilley [496]. Nickel silicides were also synthesized by Baroduc-Müller [444]. The first experiments to work out the phase diagram of the Ni–Si system were carried out by Guertler and Tammann [582] with the aid of thermal analysis and microscopic examination of alloys. The nickel contained 1.9% Co and 0.5% Fe, while the silicon was 98.9% pure. The results they obtained were then revised by Guertler [551], who reported the following nickel silicides: Ni_3Si, Ni_2Si, Ni_3Si_2, NiSi, and $NiSi_2$. The data obtained by these investigators are the basis for the phase diagram of the Ni–Si system presented in Fig. 109. According to

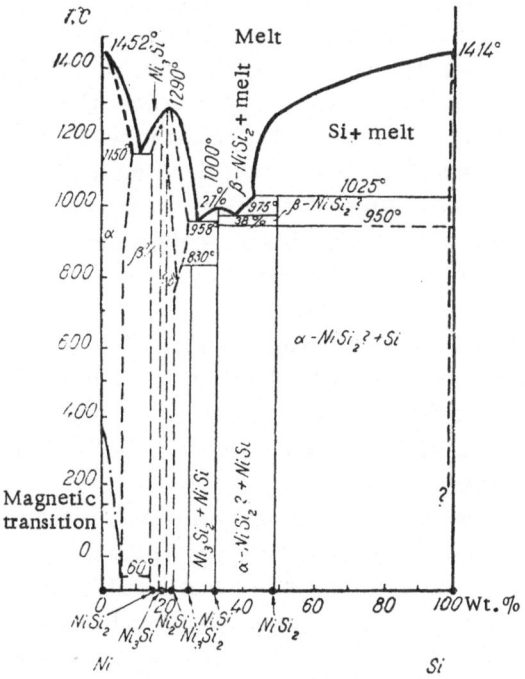

Fig. 109. Phase diagram of the Ni–Si system, according to Guertler (1917).

Guertler [551], silicon forms a solid solution in nickel up to 6% concentration at 1150°. According to Dahl and Schwartz [592], the silicon content of nickel is about 7.6% at 950°. The silicon content of the solid solution is less than 4.6% at 500°. Decomposition of the α-solid solutions proceeds very slowly [593].

An x-ray examination by Blumenthal and Hansen [28] showed that the limiting concentration of silicon in the α-solid solution is about 7.5% at 1100° and approximately 6.1% at 800°.

According to the original version of the phase diagram of the Ni–Si system (and, apparently, more accurate — A.B.), the silicide Ni_3Si is formed in the solid phase (below 1040°). According to the subsequent correction of Guertler [551], it is assumed that this compound is formed peritectically from the melt at 1250°. This phase (β) has a variable composition. On the nickel side it corresponds to the formula Ni_3Si with a structure of the Cu_3Au type [712], while on the silicon side, it corresponds to Ni_5Si_2. The thermal effect at 1120° in this region of the Ni–Si system has not yet been explained. It is possible that the β-phase is obtained as a result of a modification change of Ni_2Si (1214°), which could be present in alloys as a nonequilibrium phase. The temperatures of magnetic conversions of solid solutions of silicon in nickel have been corrected according to Kussmann and Scharnow [591].

The Ni_2Si phase forms a solid solution with excess silicon (see Fig. 101) from which a new phase, denoted as Ni_3Si_2, separates at 830° with a large heat effect. Considering the structure of the Mn–Si and Fe–Si systems, it can be presumed that this is Ni_5Si_3 (22.31% Si). This hypothesis is also confirmed by its hexagonal structure (see Table 2).

Nickel monosilicide, NiSi, melts with a straightforward maximum at 1000°, while the disilicide, $NiSi_2$, melts incongruently at 1025° and has a modification change at 950°.

Nickel forms solid solutions in silicon [497], but they have been studied very little as yet.

The Ni–Si system and the structures of nickel silicides were studied by Osawa and Akamoto [594] and others [713]. The Japanese investigators established that Ni_3Si has a cubic lattice (superlattice similar to Cu_3Au) and Ni_5Si_2 has a hexagonal lattice.

Hence it follows that the conclusions of Guertler that Ni_5Si_2 is the limiting solid solution of the same type as Ni_3Si requires re-examination. However, there is insufficient experimental material

'or this as yet. These investigators also established that the silicide
Ni₂Si has a conversion at 1214°, when the high-temperature θ-form
changes into the low-temperature δ-Ni₂Si. The temperature of the
conversion is lowered to 806° in the presence of silicon in the solid

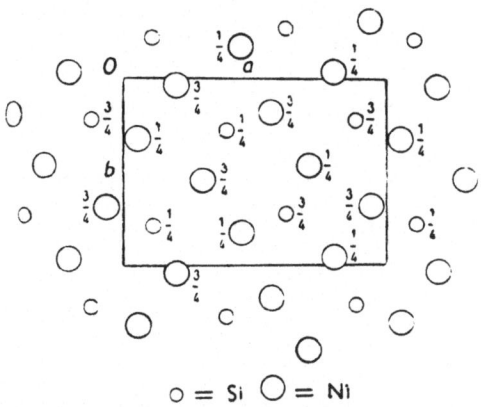

o = Si O = Ni

Fig. 110. Projection of the structure of δ-Ni₂Si
on the (001) plane, according to Toman (1952).

solution in the θ-phase. The structures of θ- and δ-Ni₂Si were determined more precisely in the Czechoslovakian Institute of Metals
by Toman [595] by an investigation of monocrystals of these compounds (for θ-Ni₂Si — on samples containing 24% Si and quenched

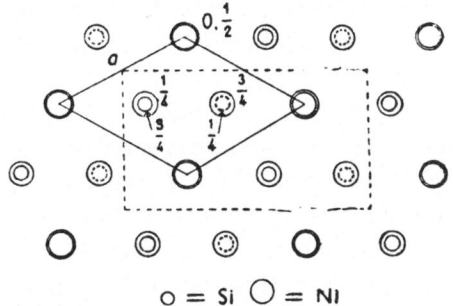

o = Si O = Ni

Fig. 111. Projection of the θ-Ni₂Si structure
on the (001) plane, according to Toman (1952).

from 964°). The constants of the elementary cells thus established
are given in Table 2. Figures 110 and 111 show projections of the
structures of θ- and δ-Ni₂Si, which show that these structures differ
primarily in the disposition of silicon atoms. As in the low-temper-

ature modification of Co_2Si, in δ-Ni_2Si the silicon atoms form zig-zag chains. In θ-Ni_2Si the silicon atoms are arranged so that in the [001] direction they are projected onto part of the nickel atoms (see Fig. 111). The distances between the centers of gravity of the atoms in both modifications of Ni_2Si are presented in Table 32.

Table 32

Distance Between Centers of Gravity of Atoms in Ni_2Si [595]

Atoms	Distance, A	
	δ-Ni_2Si	θ-Ni_2Si
Ni—Ni	2.54; 2.62; 2.72	2.45; 2.52
Si —Si	3.72	3.29
Ni—Si	2.28; 2.31; 2.34; 2.35; 2.41; 2.56; 2.62; 2.72	2.23; 2.45; 2.52

This table shows that the structures of δ- and θ-Ni_2Si differ also in the interatomic distances. In the first modification these distances are greater than in the second, for example, by approximately 13% for Si-Si. In δ-Ni_2Si, each nickel atom is surrounded by eight closest silicon atoms at different distances, while in the θ-form there are only three different Ni—Si distances. The specific gravity (pyknometric) of δ-Ni_2Si is 7.23 and that of θ-Ni_2Si, 6.85, considerably less than those calculated from the lattice constants (see Table 2). This is explained by changes in the elementary cell constants with a considerable excess of silicon (for example, for θ-Ni_2Si at a composition close to Ni_3Si_2).

The structure of the silicide Ni_3Si_2 was studied by Osawa and Okamoto [594], whose results are presented in Table 2.

Nickel monosilicide, NiSi, was described by Boren [497] as cubic of the FeSi type with $a = 4.437$ A. Osawa and Okamoto [594] considered it to be tetragonal ($a = 7.655$ A, $c = 8.45$ A, $z = 20$). However, the detailed investigation of Toman [596] showed that nickel monosilicide belongs to the rhombic system (see Table 2), MnP type, and its structure is presented in Fig. 112 and its projections in Fig. 113. This structure is a deformed NiAs type. Here the silicon atoms form zigzag chains, parallel to the y axis. They are surrounded by six nickel atoms in the form of an irregular trigonal prism and each nickel atom is in the center of an irregular octahedron of silicon atoms. The interatomic distances in NiSi are as follows: Ni–Ni–2.66; 2.69 A; Si–Si–2.58 A; Ni–Si–2.29; 2.30; 2.38; 2.44 A.

Nickel disilicide, $NiSi_2$, is known to have two modifications, which were investigated by Osawa and Okamoto [594]. The low-temperature α-$NiSi_2$ is trigonal and the high-temperature β-$NiSi_2$ is cubic, iso-morphous with one of the $CoSi_2$ modi-fications (see Table 2).

The heats of formation of Ni_2Si and NiSi were determined by Oelsen and Samson-Himmelstjerna [597] and that of Ni_3Si by Kubaschewsky [429] (see Table 2). The heats of formation of nickel silicides are similar to those of cobalt silicides, but the melting points

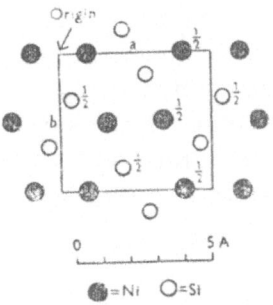

●=Ni ○=Si

Projection along the c axis

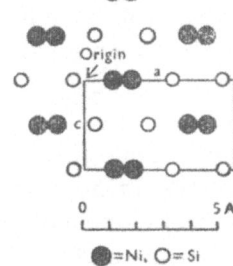

●=Ni, ○=Si

Projection along the b axis

Fig. 113. Projections of the NiSi structure along the c and b axes, according to Toman (1951).

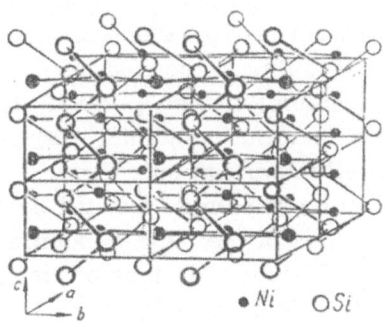

●Ni ○Si

Fig. 112. Crystal lattice of NiSi (MnP type).

are considerably lower, indicating a weakening of the bonds in the crystals of nickel silicides. The microhardness of Ni–Si alloys was studied by Savitskii and Baron [633].

The properties of nickel silicides have been studied very little as yet. The silicide Ni_2Si reacts with fluorine at normal temper-ature with ignition and with chlorine, only at red heat. This silicide is oxidized when heated in air or in oxygen. Gaseous HF, HCl, HBr, and HI decompose dinickel silicide at red heat with the formation of nickel and silicon halides. Water vapor decomposes it under the same conditions. Hydrofluoric acid dissolves this compound very readily and the other acids, with difficulty. Dinickel silicide dissolves in a mixture of HCl and HNO$_3$. Aqueous solutions of alkalies do not react with it, while molten alkalies decompose it with the formation of soluble alkali silicates and a precipitate of nickel oxides. The action of K$_2$CO$_3$ and KNO$_3$ mixture on this silicide is analogous but occurs at lower temperatures [23].

The chemical properties of other nickel silicides have not been studied.

Solid solutions of nickel in silicon are used for the manufacture of detectors and semiconductor amplifiers [42]. Nickel silicides have no practical use as yet.

Systems Formed by Silicon with Metals of the Palladium Subgroup

The atomic radii of ruthenium (1.30), rhodium (1.34), and palladium (1.37) are greater than those of the metals of the iron subgroup and this provides the geometric prerequisites for the formation of more complex silicides. The structures of the electron shells are characterized by the completion of filling of the N4d level and the beginning of filling (apart from palladium) of the O$_5$s level. The five elements following them (silver, cadmium, indium, tin, and antimony) do not form silicides and tellurium and iodine give only unstable compounds with silicon. It can be presumed that the heats of formation and melting points of silicides of the metals examined should decrease from ruthenium to palladium. The lack of appropriate thermodynamic data on the silicides of palladium group metals and of phase diagrams of the Ru–Si and Rh–Si systems make it impossible to develop in detail the rules existing here. Judging by the phase diagram of the Pd–Si system, the melting points of ruthenium and rhodium silicides should be relatively low (hardly higher than 1400–1500°). All the ruthenium, rhodium, and palladium silicides studied are formed with a decrease in volume (see Table 2).

Ru–Si SYSTEM. Moissan and Manchot [598] obtained the ruthenium silicide RuSi by fusing ruthenium and silicon in an electric furnace. The charge contained a large amount of silicon (about 82%). For isolation of the RuSi, the coarse-ground alloy was treated with sodium hydroxide and then a mixture of HF and HNO$_3$. Moissan reported that RuSi may also be obtained with the aid of copper silicides by Lebeau's method.

Buddery and Welch [599] synthesized ruthenium silicides from silicon and ruthenium in an evacuated quartz tube at 1000-1200°. The preparations obtained were then investigated by means of x-ray powder diagrams. As a result of this work, it was found that three ruthenium silicides exist: Ru_3Si_2 (structure not established), two cubic modifications of RuSi, and tetragonal Ru_2Si_3 (see Table 2). The phase diagram of the Ru-Si system has not been established.

Ruthenium monosilicide, RuSi, crystallizes in the form of white prisms with pyramidal ends and a strong metallic luster. Its specific gravity (5.4), reported by Hönigschmid [23] and then included in all textbooks, is incorrect and from the size of the elementary cell (see Table 2) it is 7.83 for one modification and 8.30 g/cc for the other. The temperature of the modification change of this compound is not known.

Ruthenium monosilicide burns in fluorine at room temperature. The reaction with chlorine proceeds at red heat. It burns in oxygen on heating. This silicide does not react with any acid. A mixture of HF and HNO_3 slowly decomposes it only on prolonged heating.

Molten caustic alkalies decompose ruthenium monosilicide with greater difficulty than pure ruthenium. The salt $KHSO_4$ in the molten state slowly decomposes it with the formation of perruthenates. Potassium hypochlorite does not react with this silicide at all, while metallic ruthenium dissolves in it readily [23].

The properties of the other ruthenium silicides have not been studied yet. They have no practical use.

Rh-Si SYSTEM. The phase diagram of this system has not been established yet. Rhodium silicides were obtained from rhodium and silicon similarly to ruthenium silicides by Buddery and Welch [599], apparently for the first time. In the Rh-Si system they found only the following intermediate phases: a rhodium-rich unidentified phase, Rh_3Si_2, RhSi, and Rh_2Si_3 or $RhSi_2$ (they are both given in Table 2). The crystal structure of only RhSi among all the rhodium silicides has been studied [600] and this was found to be the same as that of FeSi (see Table 2). Geller and Wood [600] investigated alloys of the Rh-Si system containing 50, 56, 60, and 75 at.% Si and found in them only the two phases RhSi and silicon. At 50% Si, the x-ray patterns of the alloy also showed the lines of a phase richer in rhodium than rhodium monosilicide. The interatomic distances in RhSi, found by Geller and Wood, are presented in Table 33.

Table 33

Interatomic Distances in RhSi [600]

Atom	Neighboring atom	Number of equivalent neighbors	Distance, A
Rh	Si	1	2.46±0.08
Rh	Si	3	2.44±0.05
Rh	Si	3	2.57±0.05
Rh	Rh	6	2.87±0.01
Si	Si	6	2.90±0.02

In the structure of rhodium monosilicide, the position of the silicon atoms is very similar to that in FeSi, while the parameters of the positions of the rhodium atoms differ from those of iron.

No transition into a superconducting state was observed for this silicide down to 1.3°K.

The other properties of the monosilicide and of other rhodium silicides have not been studied yet. They have no practical use.

Pd–Si SYSTEM. The Pd–Si system has been studied much more fully than the Ru–Si and Rh–Si systems. Lebeau and Jolibois [601] found that palladium reacts with silicon at relatively low temperatures (about 600°) to form two silicides, Pd_2Si, and PdSi. These investigators used thermal analysis to determine the melting points of Pd_2Si (1400°) and PdSi (900°). According to Lebeau and Jolibois, eutectic points occur at the following temperatures: for Pd–Pd_2Si (about 6% Si) — 670°; Pd_2Si–PdSi (about 16% Si) — 750°; and PdSi–Si (25% Si) — 825°. According to a microscopic investigation, alloys containing 11.7% Si (Pd_2Si) and 21% Si (PdSi) consisted of one phase. Palladium monosilicide was isolated from an alloy containing about 60% Si after treatment with potassium hydroxide solution, which dissolved the free silicon and hardly reacted with the palladium silicide.

In alloys containing not more than 20% Si, an exothermal reaction is observed in the solid phase at 600° and no satisfactory explanation has been found for this.

On the basis of these data, Hansen [28] constructed a preliminary phase diagram for the Pd–Si system.

Grigor'ev, Strunina, and Adamova [602] used thermal analysis to work out the phase diagram of the Pd–Si system. The palladium they used contained 0.95% Fe. Alloys were investigated over the range 2.0–54.5 wt. % Si. The results they obtained were presented in Fig. 114, from which it follows that the data of Lebeau and Jolibois were largely confirmed.

Solid solutions of silicon in palladium and palladium in silicon and also the limits of the fields of homogeneity of the silicides

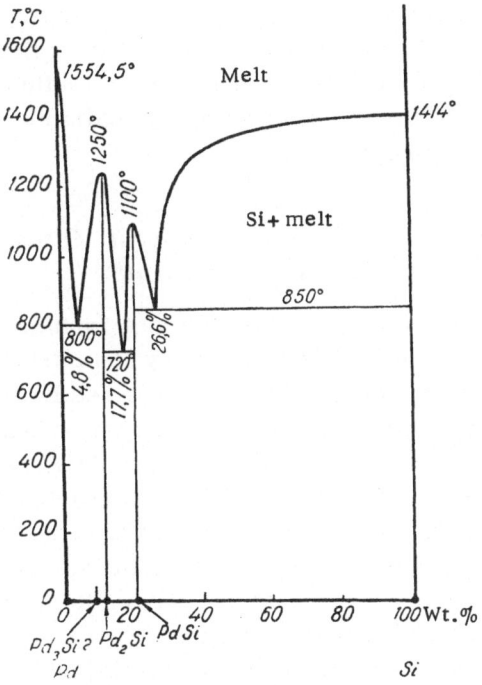

Fig. 114. Phase diagram of the Pd—Si system, according to Grigor'ev, Strunina, and Adamova (1952).

have not been investigated. The high friability of palladium–silicon alloys has been reported.

The crystal structure of PdSi was studied by Pfister and Schubert [587, 603] and found to be rhombic of the MnP type (see Table 2).

The structure of Pd_2Si (see Table 2), obtained at 700°, was found to be hexagonal [604, 605]. The valence electrons in this silicide are arranged parallel to (001). This phase has no modification changes. Anderko and Schubert [605] found that there is a low symmetry phase in the Pd–Si system with the composition Pd_3Si. According to the phase diagram of the Pd–Si system (see Fig. 113), this compound should melt incongruently and may decompose in the solid phase at 600°, causing the thermal effect that was not explained by Lebeau and Jolibois. Further experimental investigations are required to elucidate this.

Palladium monosilicide, PdSi, has a blue-gray color. When heated to red heat in oxygen, this silicide is oxidized at the surface. Hydrochloric and sulfuric acids do not react with it. This silicide dissolves in HNO_3. Aqueous solutions of caustic alkalies decompose it with the formation of metallic palladium and alkali silicates [23]. The properties of the other palladium silicates have not been studied. The Pd–Si system and palladium silicides have not found practical use yet.

Systems Formed by Silicon with Metals of the Platinum Subgroup

Available data show that the systems formed by silicon with metals of the platinum subgroup are very reminiscent of the Ru–Si, Rh–Si, and Pd–Si systems. This is explained by the similarity of the atomic radii of the metals and a certain resemblance in the structure of the electron shells (completion of filling of the d-level).

Due to the lack of data on the phase diagrams of the Os–Si and Ir–Si systems, it is impossible to make any generalizations on the systems examined. The stability of the silicides formed in them is apparently similar to that of silicides of palladium subgroup metals.

Os–Si SYSTEM. The phase diagram of the Os–Si system has not been worked out yet. Buddery and Welch [599] obtained the following osmium silicides from osmium and silicon in vacuum at $1000-1200°$: tetragonal Os_2Si_3, isomorphous with Ru_2Si_3 (see Table 2) and the disilicide $OsSi_2$, whose structure has not been established yet. Cubic OsSi also exists. The properties of osmium silicides have not been studied.

Ir–Si SYSTEM. The phase diagram of the Ir–Si system has not been established yet. The following iridium silicides were obtained only recently [599]: an unidentified iridium-rich phase, hexagonal Ir_3Si_2 (see Table 2) and IrSi, Ir_2Si_3, and $IrSi_2$, whose structures have not been determined. The silicide IrSi belongs to the rhombic system (see Table 2). The properties of iridium silicides have not been studied yet.

Pt–Si SYSTEM. In the middle of the Nineteenth Century it was known that platinum reacts readily with silicon on heating. Berzelius, Collet-Descotils, Boussingault, Winkler, Guyard, and Colson studied the interaction of platinum and silicon [23]. More or less pure platinum silicides were synthesized by various methods, including heating platinum with SiO_2 in the presence of a reducing agent.

On the basis of indirect and sometimes very inadequate criteria, the silicides obtained were assigned the formulas: $PtSi_8$ (Winkler, 1864), Pt_3Si_2, and Pt_4Si_3 (Guyard, 1876), Pt_3Si_2 and Pt_2Si (Colson, 1882), Pt_9Si, etc. Vigouroux [120] obtained pure platinum silicide Pt_2Si from a mixture of silicon and platinum in a Moissan tubular electric furnace. Lebeau and Novitzky [606] synthesized the silicide PtSi by heating in a carbon crucible briquettes pressed from sponge platinum and finely ground silicon. They reported that the formation of this compound, which is the silicide richest in silicon in the Pt–Si system, proceeds with great evolution of heat. In another experiment on the synthesis of platinum monosilicide, silver was added to reduce the melting point of the reaction mixture. In this case prismatic crystals of PtSi were obtained and these melted at 1100° (apparently, this was not completely pure PtSi — A.B.). Vigouroux [607] repeated Lebeau's experiments on the synthesis of PtSi and confirmed his results.

The early work on the synthesis of platinum silicides was examined in detail by Baroduc-Müller [444]. This work is now only of historical interest.

Voronov [608] investigated the Pt–Si system by means of thermal analysis, microscopy, and hardness and electrical resistance measurements. Platinum of 99.8% purity and very pure silicon (Kahlbaum) were used for this purpose and were melted in a graphite crucible. As a result of these investigations the phase diagram of the Pt–Si system given in Fig. 115 was constructed. It was established that there are three silicides in this system: Pt_5Si_2, Pt_2Si, and PtSi, which is analogous to the Pd–Si system. Considering this analogy, we presume that the silicide Pt_5Si_2 (4.58% Si) actually has the composition Pt_3Si (5.44% Si), similar to Pd_3Si. From Fig. 115 it follows that the region of homogeneity of the silicide Pt_2Si extends approximately to the composition (39.1 at.% Si) corresponding to the formula Pt_3Si_2. Diplatinum silicide, Pt_2Si, has a modification change at 700°. Silicon forms solid solutions in platinum up to a concentration of 0.2 wt.% Si. The formation of solid solutions of platinum in silicon has not been reported. The rest of the phase diagram of the Pt–Si system is clear from Fig. 115. It is interesting to note that the melting points of palladium and platinum silicides do not show a regular change (Pd_2Si melts at a higher temperature than Pt_2Si; and PdSi, at a lower temperature than PtSi).

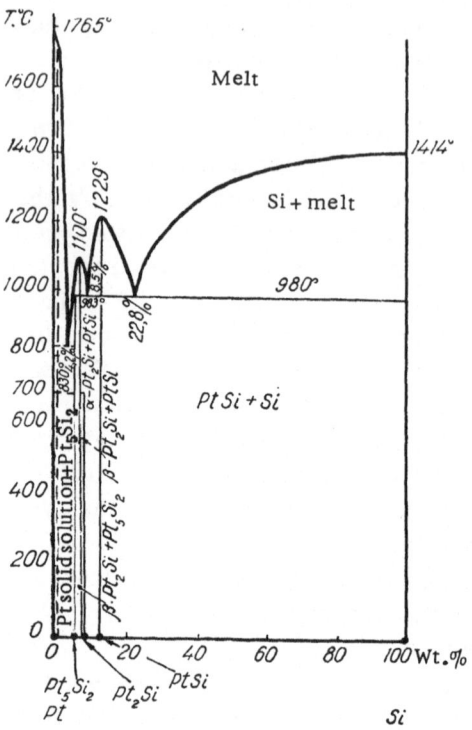

Fig. 115. Phase diagram of the Pt—Si system, according to Voronov (1936).

Buddery and Welch [599] made an x-ray investigation of platinum—silicon alloys, obtained at 1000–1200°, and found a platinum-rich phase, which they did not identify, Pt_2Si or Pt_3Si_2, an unidentified intermediate phase and PtSi.

The crystal structure of Pt_5Si_3 and the other properties of this silicide have not been studied yet. The structure of Pt_2Si (apparently the low-temperature modification) is tetragonal (see Table 2). Schubert and Anderko [604] pointed out that the elementary cell constants found by Buddery and Welch [599] actually refer to the intralattice cell. It can be assumed that the second modification of Pt_2Si will have a hexagonal structure similar to Pd_2Si (see Table 2).

The structure of the monosilicide PtSi was determined by Pfister and Schubert [587, 603]. This silicide was found to be rhombic and of the same type as PdSi (see Table 2).

All the platinum silicides are low-melting and very friable. They can be formed in a reducing atmosphere at the contact of platinum with silicon-containing materials (procelain, chamotte). It is therefore advantageous to wind platinum heaters on materials not containing SiO_2 [608], for example, articles of alumina or zirconium dioxide. When heated with SiO_2 in the presence of hydrogen, platinum evaporates due to the formation of silicides [609], indicating the appreciable volatility of these compounds.

The silicide Pt_2Si is white in color. The pyknometric specific gravity of this silicide [23] presented in Table 2 is apparently somewhat low and due to this the calculated change in volume during the formation of this compound from free silicon and platinum has an unexpected positive sign.

Chlorine reacts with Pt_2Si to form $SiCl_4$ and a residue, which dissolves in a mixture of HCl and HNO_3. The silicides Pt_2Si and PtSi are soluble only in mixtures of these acids and this is used for their chemical analysis. Platinum monosilicide, PtSi, is decomposed by molten caustic alkalies, sodium peroxide, and alkali carbonates and nitrates. When fused with tin without the addition of silicon, this silicide decomposes with the liberation of free silicon and the formation of Pt_2Si. Platinum monosilicide also behaves similarly when fused with aluminum. Molten copper decomposes it completely, as was observed by Vigouroux. It is possible that a silicide with the composition Cu_2PtSi is thus formed [23].

Platinum silicides are not used for practical purposes. As indicated above, their formation has a negative value, causing wear of platinum parts.

Systems Formed by Silicon with Lanthanides

The study of systems formed by silicon with lanthanides is only just beginning. Only the phase diagram of part of the Ce–Si system is available.

In recent years, disilicides of lanthanum, cerium, praseodymium, neodymium, and samarium have been prepared from the oxides and silicon and their crystal structures studied and shown to be isomorphous with $ThSi_2$ [610–613]. All these compounds are formed with a decrease in volume (see Table 2). It can be considered that the rest of the rare earth elements, whose electron structures and atomic radii are similar, form at least the disilicides $MeSi_2$ (with

24.30% Si for lutecium) with an increase in the melting point from LaSi$_2$ to LuSi$_2$. It is also possible that these elements, like cerium, form the monosilicides MeSi. Silicides of lanthanides of more complex composition, especially with a higher metal content, are unknown as yet but are apparently possible.

The properties of lanthanide silicides, with the exception of a few data on cerium silicides, have not been studied at all yet.

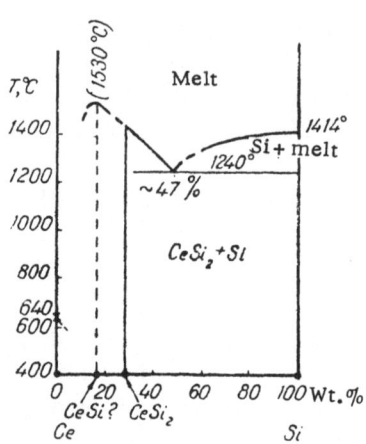

Fig.116. Phase diagram of the Ce—Si system, according to Vogel (1913).

La–Si SYSTEM. Only the disilicide LaSi$_2$ [611, 612], which is isomorphous with ThSi$_2$, has been obtained as yet in the La–Si system. The latest data of Brauer and Haag [612] on the elementary cell constants of LaSi$_2$ are presented in Table 2. Bertaut and Blum [611] obtained the values $a=4.37$ A, $c=13.56$ A, and a specific gravity of 4.95, which differ little from the values of Brauer and Haag. The properties of LaSi$_2$ have not been studied yet. The formation of other silicides in the La–Si system is probable, especially with a lower silicon content (LaSi etc.).

Ce–Si SYSTEM. In 1865 Ullick obtained the cerium silicide CeSi during the isolation of metallic cerium by electrolysis of its fluoride. Sterba [614] obtained cerium disilicide by the action of silicon on cerium oxide at high temperatures in an electric furnace (Moissan type). This silicide was obtained recently by Zachariasen [610] and also by Brauer and Haag [613] from CeO$_2$ and silicon. Vogel [615] attempted to work out the phase diagram of the Ce–Si system. The cerium he used was apparently of low purity (judging by his investigations of the Ce–Sn system, it contained 93.5% Ce). Alloys containing from 30 to 100% Si were investigated. Using thermal analysis, Vogel found only one compound — CeSi. The results he obtained are presented on the schematic phase diagram of the Ce–Si system (Fig. 116). It is obvious that this phase diagram requires further study with the use of modern investigation methods. Silicides of the composition CeSi$_{0.35}$, CeSi$_{0.5}$, and CeSi$_{0.75}$ were also found recently [689].

The structure of $CeSi_2$ was found [610] to be of the same type as that of $ThSi_2$ (see Table 2) with the following interatomic distances: Ce–Si–3.13 A, Si–Si–2.36 A.

The structure of cerium monosilicide, CeSi, has not been studied yet.

Cerium silicides are a steel gray color; they are very friable and readily ground to a black powder; they are insoluble in water, but are decomposed by its prolonged action in the presence of air. Organic liquids do not react with them. Hydrogen reduces them. Cerium silicide ignites at room temperature in fluorine and in chlorine on heating. It does not change in air, but is readily oxidized when heated to red heat. Cerium silicide reacts with sulfur and selenium at their boiling points. Cerium disilicide, $CeSi_2$, reacts with nitrogen at high temperatures [689].

Cerium silicide forms magnesium silicide when heated with magnesium. Gaseous hydrogen chloride reacts with cerium silicide at red heat with a weak glow. Acids react with it to liberate hydrogen. This compound is converted into the sulfide in hydrogen sulfide. Solutions of caustic alkalies do not react with it, while molten alkalies readily decompose it.

Cerium silicides are not used directly. However, polycomponent systems, for example, Al–Ce–Si and Al–Fe–Ce–Si, containing small amounts of cerium, are of value in technology. The addition of cerium to aluminum alloys containing iron and silicon promotes reduction of the grain size of iron silicides.

Pr–Si SYSTEM. Brauer and Haag [612, 613] prepared praseodymium disilicide, $PrSi_2$, and studied its structure (see Table 2). Its properties were not studied.

Nd–Si SYSTEM. The disilicide $NdSi_2$ has been synthesized [612, 613]. Its structure is the same as that of $ThSi_2$ (see Table 2). The properties of $NdSi_2$ have not been studied.

Pm–Si SYSTEM. This system has not been studied at all. No promethium silicides have been prepared yet.

Sm–Si SYSTEM. Samarium disilicide, $SmSi_2$ [612, 613] is isotopic with $ThSi_2$ (see Table 2). Its other properties have not been studied.

SYSTEMS FORMED BY SILICON WITH EUROPIUM, GADOLINIUM, TERBIUM, DYSPROSIUM, HOLMIUM, ERBIUM, THULIUM, YTTERBIUM, AND LUTECIUM. These systems have not been studied at all and no silicides of these elements have been prepared as yet. As mentioned above, silicides should apparently exist, especially of the type MeSi$_2$ and MeSi. The composition of the ytterbium silicide obtained by Brauer [613] was not determined.

Systems Formed by Silicon with Actinides

With the exception of actinium, all the actinides are characterized by filling of the 5f level of the electron shell and this determines the similarity of their physicochemical properties. Apart from data on the U–Si system and isolated information on thorium, neptunium, and plutonium silicides, there are no data on the systems formed by 5f elements and silicon. It is thus impossible to indicate general rules. The large and comparatively similar atomic radii of these elements in metallic and covalent bonds [620] must determine the complexity of the phase diagram structures of the silicide systems, especially in the regions poor in silicon. The phase diagram of the U–Si system is an example. At the same time, the silicon-rich regions must have a simple structure as the structure of the silicides in these systems is determined primarily by the type of packing of the metal atoms. This is also confirmed by available experimental data.

Ac–Si SYSTEM. The Ac–Si system has not been studied at all. Actinium silicides, which should exist, have not been prepared.

Th–Si SYSTEM. Thorium silicide was first obtained by Hönigschmid [486, 621, 622] in an electric furnace by reducing ThO$_2$ with silicon. It was then difficult to analyze thorium silicide due to the presence of unreduced oxide. Therefore, Hönigschmid proposed the preparation of thorium silicide from metallic thorium and silica in the presence of molten aluminum. The fusion was carried out in a porcelain crucible in vacuum at 1000°. The silicide obtained had the composition ThSi$_2$. Wedekind (1913) prepared this compound by hot-pressing of the components at 1000–1200° [27]. Hönigschmid's synthesis for the preparation of thorium disilicide was repeated by Brauer and Mitius [623], who wished to investigate the structure of the substance. They used 99.5% pure thorium and silicon in the form of silumin with 13% Si and these were melted in vacuum at 1100° in an alumina crucible. The monosilicide ThSi and other silicides with a greater thorium content probably exist but this has not been investigated yet.

The melting points of thorium silicides have not been determined yet but they will hardly be very high (hypothetically they should be at about 1600°).

Brauer and Mitius [623] studied the crystal structure of $ThSi_2$ and they also described a method for analyzing it chemically. The results of this investigation are given in Table 2 and in Fig. 117. It was found that $ThSi_2$ is not isomorphous with $TiSi_2$ and $ZrSi_2$ or with the disilicides of Group V and VI transition metals, but is a special type. In this silicide, each silicon atom is surrounded by three silicon atoms, which are arranged identically at a distance of 2.39 A, and by six thorium atoms, which are not arranged similarly and are at a distance of 3.16 A (two in one position and four in the other). Each thorium atom is surrounded by eight thorium atoms (four at a distance of

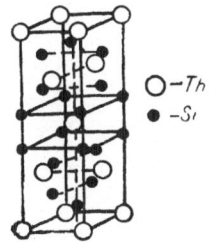

Fig. 117. Crystal lattice of α-$ThSi_2$, according to Brauer and Mitius (1942).

4.13 A and four at a distance of 4.14 A) and 12 silicon atoms at a distance of 3.15 A. There is no hexagonal packing of the atoms in this structure.

The silicon atoms in thorium disilicide form zigzag chains which are aligned one against the other and this results in the low hardness of this silicide (see data according to Cerwenka [27] in Table 2). As the distance between the silicon atoms in a chain is the same as that between the chains, a three-dimensional framework is formed from these atoms and the thorium atoms are arranged in the hollows of this framework. Analysis of the interatomic distances indicates that $ThSi_2$ has a structure that is intermediate between transition metal disilicides and $CoSi_2$, in which nonpolar bonds predominate. The heat of formation of $ThSi_2$ was determined by Robins and Jenkins [624]. Recently Th_3Si_2, $ThSi$, and β-$ThSi_2$ were also prepared and investigated [696] (see Table 2).

Thorium disilicide has a metallic, graphitelike appearance. It is not reduced by hydrogen, but burns when heated in fluorine and oxygen and sulfur and selenium vapors and also when heated to red heat in chlorine and HCl. It does not change in dry air. Dilute hydrochloric acid dissolves it slowly in the cold and rapidly when heated. Nitric acid has a slow effect on this silicide. Dilute sulfuric acid reacts with it similarly to hydrochloric acid, while concentrated sulfuric acid has only a weak effect. Dilute caustic alkali solutions do not decompose thorium disilicide, but molten KOH reacts with it rapidly with ignition. Molten $KHSO_4$ dissolves it slowly at red heat. For analysis, thorium disilicide is dissolved by treatment with a mixture of $HNO_3 + HCl$ or $HNO_3 + HF$ [23]. The other

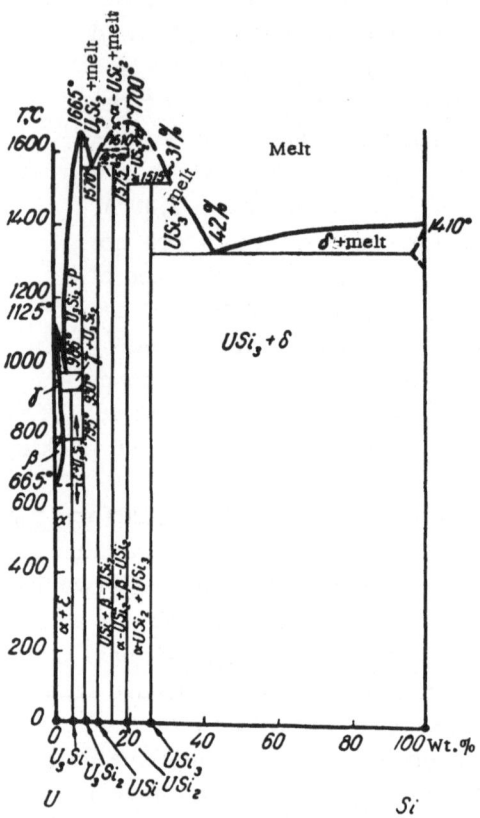

Fig. 118. Phase diagram of the U—Si system, according to Gordon, Cullity (1943), Kohn (1944), Cullity, Bitsianes, and Kaufmann (1944) with corrections according to Zachariasen (1949).

thorium silicides have properties that are similar to those of $ThSi_2$ [696]. Recent investigations [522] on the oxidizability of $ThSi_2$ (see above) showed that its change in weight at 1200° was almost a factor of 180 greater than that of $MoSi_2$ and a factor of 1.8 greater than $ZrSi_2$. Thus, thorium disilicide cannot be used as a heating element.

The Th–Si system and thorium silicides do not have any practical use yet.

Pa–Si SYSTEM. Proactinium silicides have not been synthesized, though evidently they should exist. The Pa–Si system has not been studied as yet.

U–Si SYSTEM. Uranium silicide, USi_2, was first prepared by Defacqz [625] by aluminothermal reduction of a U_3O_8 and SiO_2 mixture in a chamotte crucible in the presence of sulfur. Uranium disilicide was isolated from the melt by treating it with 10% HCl and KOH. In repeating Defacqz's experiments, Brauer and Haag [626] did not get the same results, but they did synthesize USi_2 from the elements in the presence of molten aluminum.

The phase diagram of the U–Si system (Fig. 118) was worked out in the Massachusetts Institute of Technology by Gordon, Cullity (1943), Kohn (1944), Cullity, Bitsianes, Kaufmann (1944), and Bostian using thermal analysis, microscopy, and an x-ray method [627]. These investigators found the following intermediate phases in the U–Si system: U_5Si_3, USi, U_2Si_3, USi_2, USi_3, and $U_{10}Si_3$, but they did not study the structures of these phases. The structure of uranium disilicide was first studied by Brauer and Haag [626], who found that it was cubic with $a = 4.053$ A.

The maximum solubility of silicon in γ-U is about 0.2 wt.% (1.75 at.% Si) at 980°. The solubility of silicon in β-U is less than 0.1 wt.% (less than 1 at.% of Si). The β–γ-U transition point rises from 770 to 795° when silicon is dissolved in the uranium, but the α–β-U transition point remains the same at 660°. Uranium forms a solid solution in silicon with a low uranium concentration, but its limits have not been determined accurately.

Using x-ray analysis, Zachariasen [610] investigated the intermediate phases of the U–Si system and showed that there could be only the silicides U_3Si, U_3Si_2, USi, β- and α-USi_2 and USi_3 here. Figure 118 shows the phase diagram of the U–Si system with these refinements included [627] In the light of the latest data on Me_3Si_2 silicides of Group V and VI transition metals (see above), the accuracy of replacing the silicide U_5Si_3 by U_3Si_2, which Zachariasen based only on indirect criteria (volume per atom of uranium and silicon), should be reconfirmed experimentally and by calculation.

The silicide U_3Si is formed in the solid phase and exists at temperatures below 930° [715]. It melts congruently at 1665°. Uranium monosilicide, USi, melts incongruently at a temperature somewhat above 1570°. At a composition of $U_{40}Si_{60}$, the β-USi_2 phase melts incongruently at 1610° and α-USi_2 congruently at about 1700°. The interrelations of β- and α-USi_2 have been studied very little, as is the case for all the high-temperature regions of the phase diagram of the U–Si system in the range of 35–75 at.% Si. Uranium trisilicide, USi_3, melts incongruently at 1510°. The boundaries of the homogeneity regions of these uranium silicides have not been established accurately. The rest of the structure of this system's phase diagram may be seen in Fig. 118.

Table 34

Interatomic Distances and Valences in Uranium Silicides [610]

Silicide	Direction	Interatomic distance, A	Number of valence electrons	
			U	Si
U_3Si	U_I—4Si	3.01	U_I—4.5	2.4
	U_I—8U	3.04	U_{II}—4.6	—
	U_{II}—2Si	2.92	—	—
	U_{II}—2Si	3.17	—	—
	U_{II}—4U	3.02	—	—
	U_{II}—4U	3.04	—	—
	Si—4U	2.92	—	—
	Si—4U	3 01	—	—
	Si—4U	3.17	—	—
U_3Si_2	U_I—4Si	2.96	U_I—2.2	3,2
	U_I—8Si	3.32	U_{II}—2.3	—
	U_{II}—6Si	2.92	—	—
	U_{II}—4U	3.32	—	—
	Si—2U	2.96	—	—
	Si—6U	2.92	—	--
	Si—1Si	2.30	—	—
USi	U—7Si	2.98	1,9	3,6
	U—U	3.62	—	—
	Si—7U	2.98	--	—
	Si—2Si	2.36	—	—
α-USi_2	U—12Si	3.03	2.2	4.1
	Si—6U	3.03	—	--
	Si—3Si	2.29	—	—
β-USi_2	U—12Si	3.01	2.4	4.2
	Si—6U	3.01	—	—
	Si—3Si	2,22	—	—

The crystal structure of U_3Si was found to be pseudocubic [610] (Cu_3Au type) and tetragonal (see Table 2) with the following arrangement of the atoms:

$$(0, 0, 0), \left(\frac{1}{2}, \frac{1}{2}, \frac{1}{2} \right) +;$$

$$4\,U_I \text{ in} \pm \left(0, 0, \frac{1}{4} \right);$$

$$8\,U_{II} \text{ in} \pm (y, y + \frac{1}{2}, 0), (y + \frac{1}{2}, y, 0) \text{ with } y = 0.231 \pm 0.005;$$

$$4\,Si \text{ in} \left(0, \frac{1}{2}, \frac{1}{2} \right), \left(\frac{1}{2}, 0, \frac{1}{4} \right).$$

The interatomic distances are given in Table 34. There are no bonds between silicon atoms in this silicide. Its structure is shown in Fig. 119, which clearly shows how the silicon atoms are isolated. There is a bond between the uranium atoms in U_3Si with a valence of 5.78 at a silicon valence of 4 [610].

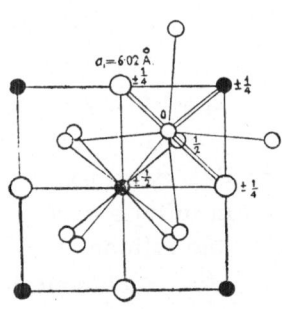

Fig. 119. Projection of the U_3Si structure normal to the tetrad axis, according to Zachariasen (1949). The large circles are Si atoms and the small ones are U in various positions.

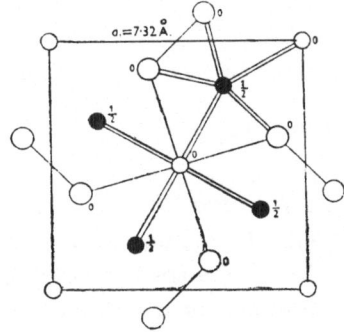

Fig. 120. Projection of the U_3Si_2 structure normal to the tetrad axis, according to Zachariasen (1949). The large circles are Si atoms and the small ones are U in various positions.

By analyzing the intensity of lines on a powder pattern and comparing the volume per atom of silicon and uranium, Zachariasen [610] came to the conclusion that the silicide U_5Si_3 is actually U_3Si_2 (see, however, our comment above on this matter). The U_3Si_2 structure is tetragonal (see Table 2). The positions of the atoms in it are as follows:

$$2 \, U_I \, in(0, 0, 0), \left(\frac{1}{2}, \frac{1}{2}, 0\right);$$

$$4 \, U_{II} \, in \pm \left(y, y + \frac{1}{2}, \frac{1}{2}\right), \left(\frac{1}{2} - y, y, \frac{1}{2}\right); \; y = 0.181 \pm 0.006;$$

$$4 \, Si \, in \pm \left(v, v + \frac{1}{2}, 0\right), \left(\frac{1}{2} - v, v, 0\right); v = 0.389.$$

The interatomic distances are given in Table 34. The structure of U_3Si_2 is characterized (Fig. 120) by the formation of pairs of silicon atoms with one covalent bond normal to the tetrad axis. In this silicide there is also a bond between uranium atoms, which have a lower valence (see Table 34).

Uranium monosilicide has rhombic symmetry (see Table 2) with the atoms in the following positions:

$$4 \, U \, in \pm \left(x_1, y_1, \frac{1}{4}\right), \left(\frac{1}{2} - x_1, y_1, +\frac{1}{2}, \frac{1}{4}\right); x_1 = 0.125 \pm 0.010,$$

$$y_1 = 0.180 \pm 0.010;$$

$$4 \, Si \, in \pm \left(x_2, y_2, \frac{1}{4}\right), \left(\frac{1}{2} - x_2, y_2 + \frac{1}{4}, \frac{1}{4}\right); \; x_2 = 0.611, y_2 = 0.028.$$

The silicide USi has a structure of the FeB type, like PtSi (see Table 2). The silicon atoms in it are joined In infinite zigzag chains along the c axis (two covalent bonds each for the silicon atoms). The Si–Si distance in its chains equals 2.36 A and the bond angle is 112°. The presence of a bond between uranium atoms in USi is doubtful [610].

An x-ray investigation of a β-USi_2 phase with the composition "U_2Si_3," showed that it belonged to the hexagonal system (see Table 2). The similarity of the volumes of the elementary cells of this phase and of α-USi_2 (52.3 $(kX)^3$ and 54.0 $(kX)^3$, respectively) indicates that this phase should be considered as β-USi_2 and not as U_2Si_3. The positions of the atoms here are as follows:

$$1 \, U \, in(0, 0, 0), 2 Si \, in \pm \left(\frac{1}{3}, \frac{2}{3}, \frac{1}{2}\right).$$

This phase is isotopic with AlB_2 and TiB_2. Each uranium atom in β-USi_2 is surrounded by 12 silicon atoms at a distance of 3.01 A and each silicon atom is attached to six uranium atoms and three silicon atoms. The interatomic distances are presented in Table 34.

The silicon atoms in β-USi$_2$ form graphitelike layers, normal to the sixfold axis of symmetry. Here there are three covalent bonds between the silicon atoms. There are no bonds between the uranium atoms and this compound differs from uranium diboride in this [628] as is seen from a comparison of the U–U distances in them:

Compound	U–U distance, A
β-USi$_2$.	3.85
UB$_2$	3.18

As is known [620], the covalent radii of uranium are as follows: $U^{4-} = 1.62$ A, $U^{5+} = 1.50$ A, $U^{6+} = 1.42$ A, and the radius of uranium in the metallic state equals 1.516 A [627].

The structure of α-USi$_2$ was found [610] to be isotopic with that of the previously investigated ThSi$_2$ (see Table 2). The positions of the atoms here are as follows:

$$(0, 0, 0), \left(\frac{1}{2}, \frac{1}{2}, \frac{1}{2} \right) + :$$

$$4 \, U \, in (0, 0, 0), \left(0, \frac{1}{2}, \frac{1}{4} \right):$$

$$8 \, Si \, in (0, 0, = z_{I}, \left(0, \frac{1}{2}, \frac{1}{4} \pm z \right); z = 0.417 \pm 0.010.$$

The interatomic distances in α-USi$_2$ are presented in Table 34. The silicon atoms are joined by covalent bonds in α-USi$_2$ to form a three-dimensional framework with the uranium atoms arranged in the hollows. This framework is interesting in that it shows the possibility in principle of the existence of a third modification of carbon — tetragonal with $a = 2.455$ A, $c = 8.50$ A, and a density 3.10 g/cm^3, in which eight carbon atoms occupy the position of the silicon atoms in α-USi$_2$.

If it is assumed that for uranium the number of valence electrons equals 2.3 (as in some other uranium compounds) with a coordination number of 12 and the radius of the uranium atom corresponding to this, equal to 1.636 A, and the number of valence electrons of silicon is the normal four, then the calculated interatomic distances in uranium silicides are in good agreement with those determined experimentally [610].

The symmetry and lattice constants of USi$_3$ were determined by Kaufmann, Cullity, and Bitsianes [627] (see Table 2). Alloys of the U–Si system, containing more than 5 wt. % (more than 30 at. %) Si are friable and difficult to polish.

Uranium disilicide, USi_2, has a metallic luster. According to Kissling [661], β-USi_2 has very great anisotropy of thermal expansion. According to x-ray data, the linear thermal expansion coefficient of this silicide along the a axis equals $+55 \cdot 10^{-6}$, and along the c axis, $-25 \cdot 10^{-6}$. Chlorine reacts with it at 500°. The oxidation of uranium disilicide in oxygen begins at 800°, while it is almost unchanged in air. This compound is very resistant to the action of acids. Hydrofluoric acid dissolves it in the cold, while a mixture of HNO_3 and HCl has no action. For chemical analysis, uranium disilicide is dissolved in a mixture of HF and HNO_3. Aqueous solutions of caustic alkalies do not decompose uranium disilicide, but molten alkalies or $KHSO_4$ do so readily. A molten mixture of alkali carbonates and nitrates decompose it only slightly.

No technical application of uranium silicides has been described in the literature.

Np–Si SYSTEM. The phase diagram of the Np–Si system has not been worked out yet. Sheft and Fried obtained neptunium disilicide $NpSi_2$ by heating NpF_3 and silicon at 1500° for five minutes in a BeO crucible in vacuum [629, 630]. The crystal structure of this silicide was found to be isotopic with α-$ThSi_2$ (see Table 2) with the Np–Si distance equal to 3.02 and Si–Si equal to 2.28 A [610]. Neptunium disilicide is a hard, metallike substance, which does not change under the action of water but rapidly reacts with hydrochloric acid. Its melting point has not been determined yet, but it can hardly be much above 1500°.

Neptunium monosilicide, NpSi, presumably exists as well.

Pu–Si SYSTEM. The phase diagram of the Pu–Si system has not been studied yet. Westrum [see 610, 629] prepared plutonium disilicide, $PuSi_2$, by heating PuF_3 and $CaSi_2$ in vacuum in a BeO crucible at 1550°. Plutonium disilicide may also be obtained by heating plutonium fluoride and silicon in a BeO crucible in vacuum $(5 \cdot 10^{-5}$ mm Hg) at 1300° for 15 minutes. The structure of the phase was found to be [610] isotopic with that of $ThSi_2$ (see Table 2) with the interatomic distances Pu–Si equal to 3.02 A and Si–Si, 2.27 A, indicating the presence of silicon covalent bonds (as with $NpSi_2$) and the absence of bonds between the plutonium atoms. With excess plutonium, corresponding approximately to the formula Pu_2Si_3 (about 15% Si), a new phase was obtained [631], namely β-$PuSi_2$ with a hexagonal system isotopic with β-USi$_2$ (see Table 2). In this phase each plutonium atom is surrounded by 12 silicon atoms at a distance of 3.03 A and each silicon atom is surrounded by six plutonium atoms and three silicon atoms with the Si–Si distance equal to 2.24 A. The phase interrelations of α- and β-PuSi$_2$ have not been studied yet.

An alloy with the composition of "PuSi$_3$" consisted only of α-PuSi$_2$ and silicon even after being annealed at 1150° for one hour [631] and this disproves the hypothesis of Frost and Maskrey [632] on the existence of plutonium trisilicide, PuSi$_3$.

It can be assumed that there should exist in the Pu–Si system silicides with a greater plutonium content than in PuSi$_2$, for example, the monosilicide PuSi (10.5% Si), which has not been described in the literature yet.

The melting points in the Pu–Si system, in particular of PuSi$_2$, have not been published yet. It can be assumed that the melting point of PuSi$_2$ will not be much above 1500°.

Plutonium disilicide has a silvery metallic luster and a high hardness and friability. When heated in air to 700°, PuSi$_2$ is oxidized to form PuO$_2$.

No technical application of plutonium silicides has been described in the literature.

SYSTEMS FORMED BY SILICON WITH AMERICIUM, CURIUM, BERKELIUM, CALIFORNIUM, EINSTEINIUM, FERMIUM, AND MENDELEVIUM. These systems have not been studied at all and no silicides of the metals indicated have been prepared yet, though, judging by the analogy with neptunium and plutonium, they should be formed.

SOME GENERALIZATIONS

The present results of the study of binary systems of silicon with elements now known are given in Table 35.

Table 35

Results of Studying Binary Systems with Silicon

Extent of study:	With chemical compounds	Without chemical compounds	Total
Phase diagrams established	21	14	35
Phase diagrams not established	32	7	39
Assumed by analogy .	26	—	26
Total	79	21	100

The table shows that silicon reacts with 79% of the elements, forming numerous chemical compounds (more than 160 of them are known at present; see Table 2). This is an indication of the high chemical activity of silicon. The characteristics of binary systems with silicon are given in Table 36 relative to the position of the elements in the Periodic Table.

One may conclude from the data in Table 36 that silicon does not react only with Group B metals, apart from copper, but does form chemical compounds with Group A metals with the exception of beryllium. Within the limits of one group, with an increase in atomic weight the structure of the phase diagrams of systems of Group B metals with silicon shows increasing immiscibility of the components in the molten state. In the systems of silicon with mercury, thallium, lead, and bismuth, the phenomenon of two immiscible melts forming is completely realized.

Excluding the Ba–Si system, which has been studied very little, one may consider that Group A metals and metalloids of Group B form few silicides, while it is characteristic of transition metals of Groups IV–VIII to have numerous compounds with silicon, sometimes with complex formulas. In these systems, silicides are formed which have metallic properties (high heat and electrical conductivity and metallic luster). Such silicides are of the general electronic type [628]. They include typical metallic compounds (for example, copper silicides). The higher the silicon content of the silicides, the weaker is the bond in them between the metal atoms and this is replaced by a strong Me–Si bond. This bond also has a metallic character [662]. The Me–Me bond is practically absent from disilicides. Silicides rich in silicon have homeopolar Si–Si bonds [467]. This change in bonds is illustrated graphically by uranium silicides which are examined above.

The thermodynamic characteristics of transition metal silicides have been studied very little as yet. Approximate data on the heats of formation of some silicides of this type according to Robins and Jenkins [662] and Brewer and Krikorian [689] are given in Table 37 with those of carbides, nitrides, and borides [670].

The table shows that the heat of formation of silicides is much less than that of nitrides. The heats of formation of silicides, borides, and carbides are similar. This is explained by the similarity of the first ionization potentials of silicon, boron, and carbon (7.39, 8.40, and 11.21 ev, respectively), while that of nitrogen is considerably greater (14.47 ev). With an increase in the silicon content of the silicides studied, the heat of formation of the latter falls. Molybdenum disilicide has a greater heat of formation than do $CrSi_2$ and WSi_2. The heats of formation of disilicides of Group IV transition metals increase with an increase in the *spd* character of the hybrid bond [624]. A similar phenomenon occurs in $TaSi_2$, but not in VSi_2 and $NbSi_2$. The bond energy is considerably lower in silicides than in carbides. The borides have a position intermediate between them [689]. An exceptionally high value is observed for the heat of formation of BaSi (-181.5 kcal/mole).

Compounds of silicon with metals have a much wider range of homogeneity than compounds with metalloids. The latter hardly form solid solutions with silicon or with the metalloids.

Among the compounds of silicon with metalloids, SiF_4 was found to have the maximum heat of formation (-370 kcal/mole). This value is in general the greatest (relative to 1 g-at. of Si) of all the heats of formation of binary silicon compounds.

The next greatest heat of formation is that of silica (-209.3 kcal/mole); this is one of the main reasons why silicon is found in

Table 36

	I A	II A	III A	IV A	V A	VI A	VII A
Period 1 (0)	H SiH$_4$ and many other						
Period 2 (2)	Li Li$_{15}$Si$_4$ Li$_2$Si	Be Eu-tec-tic					
Period 3 (2-8)	Na NaSi Na$_2$Si	Mg Mg$_2$Si					
Period 4 (2-8)	K KSi KSi$_3$?	Ca Ca$_2$Si CaSi CaSi$_2$	Sc	Ti Ti$_5$Si$_3$ TiSi TiSi$_2$	V V$_3$Si V$_2$Si? V$_5$Si$_3$ VSi$_2$	Cr Cr$_3$Si Cr$_5$Si$_3$ CrSi CrSi$_2$	Mn Mn$_3$Si Mn$_{15}$Si$_3$ MnSi MnSi$_2$
Period 5 (2-8-18)	Rb RbSi RbSi$_3$?	Sr SrSi SrSi$_2$	y YSi$_2$	Zr Zr$_2$Si Zr$_5$Si$_3$ Zr$_3$Si$_2$ Zr$_6$Si$_5$ ZrSi, ZrSi$_2$	Nb Nb$_4$Si Nb$_2$Si? Nb$_5$Si$_3$ NbSi$_2$	Mo Mo$_3$Si Mo$_5$Si$_3$ MoSi$_2$	Tc ?
Period 6 (2-8-18)	Cs CsSi CsSi$_3$?	Ba BaSi BaSi$_2$ BaSi$_3$ Ba$_5$Si$_3$? BaSi$_4$?	La—Lu (LaSi$_2$)	Hf (Hf$_5$Si$_4$?) HfSi HfSi$_2$	Ta Ta$_{4,5}$Si Ta$_2$Si Ta$_5$Si$_3$ TaSi$_2$	W (W$_5$Si?) W$_5$Si$_3$ WSi$_2$	Re Re$_3$Si ReSi ReSi$_2$
Period 7 (2-8-18-32)	Fr ?	Ra ?	Ac				
Lanthanides (2—8—18)			La LaSi$_2$	Ce CeSi CeSi$_2$	Pr PrSi$_2$	Nd NdSi$_2$	Pm ?
Actinides (2—8—18—32)			Ac ?	Th ThSi ThSi$_2$ Th$_3$Si$_2$	Pa ?	U U$_3$Si U$_3$Si$_2$ USi USi$_2$ USi$_3$	Np NpSi NpSi$_2$

Characteristics of Bin-
Relation to the Position
Periodic

ary Silicon Systems in
of the Elements in the
Table.

VIII A			I B	II B	III B	IV B	V B	VI B	VII B	0
										He
					B $B_6Si?$ $B_3Si?$	C SiC	N SiN_2 Si_2N_3 Si_3N_4 $(Si_2N_2)x$	O SiO SiO_2	F SiF_4 Si_2F_6	Ne
					Al Eutectic	Si --	P SiP	S SiS SiS_2	Cl $SiCl_4$ and many other	Ar He Not determined
Fe Fe_3Si Fe_5Si_3 FeSi $FeSi_2$	Co Co_3Si Co_2Si Co_5Si_3 CoSi $CoSi_2$ $CoSi_3$	Ni Ni_3Si Ni_5Si_2 Ni_2Si Ni_3Si_2 NiSi $NiSi_2$	Cu Cu_6Si Cu_5Si Cu_4Si $Cu_{15}Si_4$ Cu_3Si	Zn Eutectic	Ga Eutectic	Ge Solid solutions	As $SiAs_2$ SiAs	Se SiSe $SiSe_2$	Br $SiBr_4$ and many other	Kr
Ru Ru_3Si_2 RuSi Ru_2Si_3	Rh Rh_3Si_2 RhSi Rh_2Si_3 $RhSi_2$	Pd Pd_3Si Pd_2Si PdSi	Ag Eutectic	Cd Eutectic	In Eutectic	Sn Eutectic	Sb Eutectic	Te SiTe $SiTe_2$	I SiI_4 etc.	Xe
Os Os_2Si_3 OsSi	Ir Ir_3Si_2 IrSi Ir_2S_3 $IrSi_3$	Pt Pt_5Si_2 Pt_2Si PtSi	Au Eutectic	Hg He Immiscible	Tl He Immiscible	Pb He Immiscible	Bi He Immiscible	Po ?	At ?	Rn

Sm $SmSi_2$	Eu ?	Gd ?	Tb ?	Dy ?	Ho ?	Er ?	Tu ?	Ib ?	Iu ?
Pu $PuSi_2$	Am ?	Cm ?	Bk ?	Cf ?	E ?	Fm ?	Md ?		

Table 37

Heat of Formation, $\Delta H_{298}°$, of Silicides, Carbides, Borides, and Nitrides of Some Transition Metals

Group	Silicide	kcal/g-at. of metalloid	Carbides	kcal/g-at. of metalloid	Borides	kcal/g-at. of metalloid	Nitrides	kcal/g-at. of metalloid
IV	Ti_5Si_3	46	TiC	44	TiB_2	36	TiN	80
	TiSi	31			Ti_2B_3	<21		
	$TiSi_2$	16						
	Zr_2Si	50						
	Zr_6Si_5	41						
	Zr_5Si_3	46	ZrC	40	ZrB	39		
	ZrSi	35			ZrB_2	<39	ZrN	82
	$ZrSi_2$	19			ZrB_{12}	<10	HfN	78
	$ThSi_2$	21	ThC_2	22	ThB_4	<13	Th_3N_4	78
					ThB_6	<11		
V	V_2Si	37(?)	VC	28	—		VN	60
	VSi_2	<10						
	$Nb.Si_3$	22						
	$NbSi_2$	<12	NbC	19	NbB_2	<18	NbN	59
	Ta_2Si	29,3						
	Ta_5Si_3	26,7	TaC	9—38	TaB_2	<26	TaN	58
	TaSi	14						
	$TaSi_2$	11,6						
VI	$CrSi_2$	<12	Cr_4C	16	CrB_2	<15	CrN	29
			Cr_7C_3	14			CrN_2	26
	$MoSi_3$	21	Cr_3C_2	11				
	$MoSi_2$	16	Mo_2C	—4	Mo_2B	25,5	Mo_2N	17
					Mo_3B_2	21		
					MoB	16,3		
					MoB_2	11,5		
	$WSi_{0.7}$	18			Mo_2B_5	10		
	WSi_2	11	WC	8	W_2B	20—28	W_2N	17
					WB	12—22		
	$CeSi_2$	26			W_2B_3	5—9		

nature almost exclusively in the form of oxygen compounds (silica and silicates). The heat of formation of silicon disulfide (−34.7 kcal/mole) is almost a factor of six less than that of silica, which indicates the much lower stability of SiS_2 as compared with SiO_2. There is a characteristic sharp increase in the heat of formation, ΔH_{298} (in kcal/g-at. of Si), of silicon compounds from carbon to fluorine, as may be seen from the following:

$$SiC \quad Si_3N_4 \quad SiO_2 \quad SiF_4$$
$$26,5 \quad 59,8 \quad 209,3 \quad 370,0$$

In contrast to carbon, silicon does not form cyclic compounds with hydrogen or compounds with a very long chain. Under the conditions of the earth's crust, the differences between these elements appear even greater. Here almost all of the silicon is found in the form of oxygen compounds, while a considerable part of the carbon is included in the composition of organic compounds. Thus, from the present-day point of view there is hardly any basis for drawing a comparison between carbon and silicon as was attempted by many investigators at the beginning of the Twentieth Century.

Silicides of Group IV–VI transition metals are not inclusion phases like borides, carbides, and nitrides due to the considerably greater covalent radius of silicon (1.17 A) than that of boron (0.91 A), carbon (0.77 A), and nitrogen (0.71 A). As a result, the melting points of their silicides are generally lower than those of inclusion phases. This agrees with the lower thermodynamic stability of silicides as compared with that of inclusion phases. Here are the melting points of some silicides and inclusion phases:

TiC	TiN	TiB_2	Ti_5Si_3	$TiSi_2$	TaC	TaN	TaB_2	Ta_5Si_3	$TaSi_2$
3140	2950	2980	2120	1540	3800	3090	3100	2500	1560

The maximum melting points of the silicides are also considerably lower than those of inclusion phases and oxides; this may be seen from the following:

Carbide (HfC)	Nitride (HfN)	Boride (HfB_2)	Oxide (ThO_2)	Silicide (Ta_5Si_3)
3890	3310	3250	3300	2500

Processing of the data in Table 2 shows that the silicides studied (91 compounds) are distributed in the following way with respect to melting points:

Melting point, °C	Number of cases, %
Below 500	37
500–1000	14
1000–1600	27
1600–2000	10
2000–2500	10
Above 2500	0

Thus, only 20% of all the silicides studied are refractory materials (melting point above 1600°). These data show that in general the silicides are not very stable compounds. The microhardness of the inclusion phases is also much greater than that of the silicides (see Table 2). The silicides of Group IV–VI transition metals are diamagnetic in contrast to the paramagnetic inclusion phases [628].

The binary compounds of silicon have various crystal structures (see Table 2), which are sometimes very complex. The 120 structures of these compounds studied are distributed in the following way according to their crystal systems:

Crystal system	Number of cases, %
Cubic	25
Hexagonal (and trigonal)	30
Tetragonal	28
Rhombic	16
Monoclinic	1
Triclinic	0
	100

The data given indicate the predominance of high symmetry in the binary silicon compounds.

The crystal structures of the binary silicon compounds are very diverse. Thirty-seven space groups are represented in 91 of the structures that have been studied up to now (see Table 2). The space group D_{6h}^3 ($D8_8$ – Mn_5Si_3 type) is the most common (ten substances) and the next are D_{4h}^{18} (one of the Me_5Si_3 tetragonal types) (nine substances) and D_{4h}^{19} ($ThSi_2$ type). All these silicides have complex structures.

The disilicides have many types of structures which differ with respect to the presence of bonds between the silicon atoms. Their various types are determined by the ratio $R_{Si} : R_{Me}$. When the value of this ratio is considerably less than unity, complex structures are formed ($CaSi_2$, β-USi_2, $ThSi_2$). At a ratio close to unity, the structures are simpler ($TiSi_2$, $CrSi_2$, $MoSi_2$). The $ZrSi_2$ type has an intermediate position between these types. One of the $TiSi_2$ modifications (D_{2h}^{24}) may be considered as a rhombically deformed diamond structure. The hexagonal VSi_2, $NbSi_2$, and $TaSi_2$ (D_6^4) have a structure composed of Me–Si layers. The disilicides $CrSi_2$, $MoSi_2$, WSi_2, $MnSi_2$, and $ReSi_2$ have a layer structure of the CaC_2 type.

Silicides of calcium, manganese, and the iron group of metals have chain structures. A lattice of the β-W type is characteristic of some silicides with the composition Me_3Si. In the monosilicides ZrSi (FeB type) and CaSi (CrB type) there is a transition from silicon chains to metal ones (NiSi). Fe–Si groups are found in FeSi. The Si–Si distance is less in ZrSi than in $ZrSi_2$, which indicates the greater stability of the former silicide as compared with the latter. This is seen, for example, in the melting points, which are 2095° for ZrSi and 1520° for $ZrSi_2$. The melting points of the silicides depend primarily on the silicon elements in their structure [661]. The structures of the silicides of Group V–VI transition metals are very sensitive to traces of carbon and nitrogen, which stabilize the hexagonal $D8_8$ types.

The Zr_2Si, Ta_2Si, and Mg_2Si phases are polyfluorites; Ni_2Si has a deformed lattice of the NiAs type and Pd_2Si, of the Fe_2P type.

Due to the large size of the silicon atom, the molar volumes of the silicides depend less on the radius of the metal atom than, for example, in the case of carbides or nitrides. This affects the solubility of the substance in the solid phase. Thus, for example, the pairs ZrC and VC, and ZrN and VN hardly form solid solutions, while the silicides Zr_5Si_3 and V_5Si_3 ($D8_8$ type) probably form solid solutions at Zr_5Si_3 contents of 0–20 and 50–100 mol. %. Due to their less "loose" structure, $ZrSi_2$ and VSi_2 do not form solid solutions.

Table 2 shows that the great majority of binary silicon systems with metals (94%) are formed with a decrease in volume relative to the volume of the reacting substances, whereas on the great majority of those with metalloids (93%) are formed with an increase in volume. This shows that in general, silicon compounds with metals should be more stable under pressure than those with metalloids.

We may conclude from the data given that binary silicon compounds, including silicides (compounds with metals), differ from carbon compounds. They cannot be considered as typical metal phases either. Therefore, as in the case of boron compounds, the chemistry of silicon compounds is a field in itself.

SUMMARY

Progress in chemistry in the last few years has made it possible to solve numerous scientific problems and open up new possibilities in technology. The chemistry of silicon and its compounds is of great interest, especially in connection with the widespread distribution of silicon in nature and the valuable properties of its binary and other compounds.

We have tried to summarize the accumulated data on these interesting materials. A series of binary silicon compounds (silica, carborundum, and ferrosilicon) are already of great importance in the national economy. Apparently, some other silicon compounds (for example, molybdenum disilicide for manufacturing electrical heating elements or coatings) will be widely used in technology quite soon.

The review of the binary silicon systems shows that it would be most desirable to continue their study. It would be advantageous to first establish the phase diagrams of the B–Si, Hf–Si, Th–Si, N–Si, Re–Si systems and also define more accurately the structure of a series of phase diagrams that have been partly worked out (C–Si, Si–O, Co–Si, Ce–Si) and the boundaries of the homogeneity regions of transition metal silicides. The Me–Si–B systems are very important technically but they have not been studied. A study of the B–C–Si system, including establishing its phase diagram would be very valuable in the technology of refractory materials.

Determination of the thermodynamic constants of binary silicon compounds should give reliable criteria for extending their practical use. Besides this, it is necessary to investigate thoroughly the technical properties of transition metal silicides.

Of particular importance would be the study of silica and the more accurate determination of its modification changes and the dependence of these changes on temperature, pressure, and various additives. Interesting data obtained in recent years have shown that silica has still not been studied sufficiently. These data make it possible to find a new solution to a series of problems in the petrography of natural rocks and the production of technical products.

The extensive investigations being carried out on the chemistry of silicon and its compounds, both in the USSR and abroad, indicate that these gaps will soon be filled. As a result, new data will undoubtedly be obtained, which will be valuable for science and technology.

BIBLIOGRAPHY

1. H. Remy. Lehrbuch d. anorg. Chem. Vol. I (Leipzig, 1931).
2. Gmelins. Handbuch d. anorg. Chem., 8th ad., system No. 15, Silicium (Berlin) (This source was not at our disposal.—A. B.).
3. K. A. Andrianov and M. V. Sobolev. High-Molecular Organosilicon Compounds (State Defense Industry Press, 1949).
4. A. P. Kreshkov. Organosilicon Compounds in Technology (State Building Literature Press, 1950).
5. K. A. Andrianov. Organosilicon Compounds (State Chem.-Press, Moscow, 1955).
6. H. Le Chatelier. La silice et les silicates (Paris, 1914).
7. R. B. Sosman. The Properties of Silica (N.Y., 1927).
8. W. Eitel. The Physical Chemistry of the Silicates (Univ. of Chicago Press, 1954).
9. W. L. Bragg. Structure of Silicates (United Sci. Tech. Press, 1934).
10. B. S. Shvetsov. Introduction to Silicon Chemistry (Gizlegprom, Moscow, 1936).
11. V. I. Vernadskii and S. M. Kurbatov. Silicates, Aluminosilicates, and Their Analogs of the Earth (United Sci. Tech. Press, Leningrad, 1937).
12. A. I. Avgustinik. Physical Chemistry of Silicates (Leningrad, 1947).
13. V. Sobolev. Introduction to Silicate Mineralogy (L'vov State University Press, L'vov, 1949).
14. F. Hall and H. Insley. Phase Diagrams for Ceramists (Am. Ceram. Soc., 1950).
15. K. S. Evstrop'ev and N. A. Toropov. Chemistry of Silicon and Physical Chemistry of Silicates (State Building Literature Press, Moscow, 1950).
16. G. V. Kukolev. Chemistry of Silicon and Physical Chemistry of Silicates (State Building Literature Press, Moscow, 1951).
17. P. P. Budnikov and A. S. Berezhnoi. Reactions in Solid Phases (State Building Literature Press, Moscow, 1949) pp. 58-84.

.8. D. S. Belyankin, B. V. Ivanov, and V. V. Lapin. Petrography of Technical Rock (Acad. Sci. USSR Press, Moscow, 1952).

.9. D. S. Belyankin, V. V. Lapin, and N. A. Toropov. Physicochemical Systems of Silicate Technology (State Building Literature Press, Moscow, 1954).

0. O. K. Botvinkin. Physical Chemistry of Silicates (State Building Literature Press, Moscow, 1955).

1. H. Moissan. Le four electrique (Steinheil, Paris, 1897).

2. O. P. Watts. "An investigation of the borides and silicides", Bull. Univ. Wisconsin No. 145: Eng. Ser. 3, 251 (1906).

3. O. Hönigschmid. Karbide und Silizide (W. Knapp, Hall, 1914) pp. 141-257.

4. L. A. Bromley, P. W. Gilles, N. L. Lofgren, and C. M. Quill. "The chemistry and metallurgy of miscellaneous materials". Thermodynamics (McGraw-Hill Co., N.Y., 1950) pp. 40-59.

5. J. J. Harwood. "New materials developments", Materials for Product Development (Clapp and Poliak, N.Y., 1953) pp. 27-72.

3. R. Kieffer and Schwarzkopf. Hartstoffe und Hartmetalle Wien (Spring.-Verl., 1953).

7. P. Schwarzkopf and R. Kieffer. Refractory Hard Metals (Macmillan Co., N.Y., 1953).

3. M. Hansen. Aufbau der Zweistofflegierungen (Spring.-Verl., Berlin, 1936). There is a Russian translation (Metallurgy Press, Moscow, 1941).

3. F. Eisenkolb. Die Neuere Entwicklung der Pulvermetallurgie (Veb Verl. Tech., Berlin, 1955) pp. 388-397.

). E. Zintl, W. Brauning, H. L. Grube, W. King, and W. Morawietz. Z. anorg. Chem. 245, 1, 1 (1940).

.. G. Gire. Compt. rend 194, 884 (1932).

.. G. Economos. Ind. Eng. Chem. 45, 2, 458 (1953).

. G. Economos and W. D. Kingery. J. Am. Ceram. Soc. 36, 12, 403 (1953).

. M. Humenike and W. D. Kingery. J. Am. Ceram. Soc. 37, 1, 18 (1954).

'. W. D. Kingery and F. A. Halden. Bull. Am. Ceram. Soc. 34, 4, 117 (1955).

. P. Schwarzkopfe and F. W. Glaser. Z. Metallk 44, 8, 353 (1953).

. A. N. Winchell and G. Winchell. Optical Mineralogy [Russian Translation] (Foreign Lit. Press, 1953) p. 26.

. A. E. Fersman. Geochemistry IV (State Chem. Press, 1939) p. 83.

39. O. Ruff and M. Konschak. Electrochem. 32, 518 (1926).
40. L. Brewer and C. M. Quill. The Chemistry and Metallurgy of Miscellaneous Matter (N.Y., 1950) p. 13.
41. B. C. Weber and P. S. Hessinger. J. Am. Ceram. Soc. 37, 267 (1954).
42. B. I. Boltaks. J. Theo. Phys. 20, 3 (1950).
43. C. J. Gallagher. Phys. Rev. 88, 721 (1952).
44. H. Kantsky and L. Hoose. Z. Naturforsch. 8b, 45 (1953); Ber. 86, 1226 (1953).
45. E. R. Johnson and J. A. Amick. J. Appl. Phys. 25, 1204 (1954).
46. F. Heyd, F. Khol, and A. Kochanovska. Collection Czechoslov. Chem. Commun. 12, 502 (1947); Chem. Abs. 42, 5296 (1948).
47. R. Schwarz. Angew. Chem. 53, 6 (1940).
48. R. R. Ridgway and B. L. Boiley. U.S. Patent No. 2027786 (1936).
49. G. R. Finley. J. Electrochem. Soc. 99, 58 (1952).
50. P. V. Gel'd, N. N. Serebryanikov, and P. M. Sukharev. Proc. Ural Branch of the Acad. Sci. USSR 2, 244 (1956).
51. A. Stock and C. Somieski. Ber. 56, 139, 247 (1923).
52. B. N. Dolgov. Chemistry of Organosilicon Compounds (State Chem. Press, 1933).
53. F. Wöhler and H. Buff. Ann. 103, 218 (1857).
54. A. Stock and C. Somieski. Ber. 49, 144 (1916).
55. K. Clusius. Z. Phys. Chem. 23, 213 (1933).
56. B. F. Ormont. Structures of Inorganic Substances (State Tech. Theo. Press, 1950).
57. R. Spitzer, W. J. Howell, and V. Schomaker. J. Am. Chem. Soc. 64, 62 (1942).
58. M. Dyatkina. J. Phys. Chem. 20, 5 (1946).
59. N. Natta. Gazz. Chim. Ital. 60, 911 (1930).
60. P. P. Budnikov and E. A. Shilov. Chem. Ind. 8 (1926); Z. Angew. Chem. 39, 765 (1926).
61. R. Martin. J. Chem. Soc. 105, 2836 (1924).
62. G. Schwarz and G. Pietsch. Z. anorg. Chem. 232, 249 (1937).
63. K. A. Andrianov. Proc. Acad. Sci. USSR 28, 66 (1940).
64. R. Schwarz and H. Meckbach. Z. anorg. Chem. 232, 241 (1937).
65. G. Rochow and R. Didtschenko. J. Am. Chem. Soc. 74, 5545 (1952).
66. W. Schumb and C. Klein. J. Am. Chem. Soc. 59, 261 (1937).
67. J. Friedel. Jahresber., 210 (1868).
68. O. Hassel and H. Kringstad. Z. Phys. Chem. 13, 1 (1931).

69. R. Schwarz and A. Pfugmacher. Ber. 75, 1062 (1942).

70. H. Moissan. Compt. rend. 134, 1083 (1902); 135, 1284 (1902); Bull. soc. chim. 3, 29, 443 (1903).

71. W. Klemm and M. Struck. Z. anorg. Chem. 278, 117 (1955).

72. H. Nowotny and E. Scheil. Metalforschung 2, 76 (1947).

73. E. Hohmann. Z. anorg. Chem. 257, 113 (1948).

74. R. Schwarz and C. Danders. Chem. Ber. 80, 444 (1947).

75. R. Schwarz. Angew. Chem. 51, 328 (1938).

76. E. Vigouroux. Compt. rend. 122, 318 (1896).

77. P. Lebeau. Compt. rend. 141, 889 (1905).

78. C. S. Smith. J. Inst. Met., London, 40, 359 (1928); AIMME, Tech. Publ. No. 142, 25 (1928).

79. G. A. Kashchenko. Bases of Metallography (Metallurgy Press, 1949) p. 639.

80. K. Jokibe. Kinzoku-no-Kenkyu 8, 483 (1931); see J. Inst. Met., London, 47, 651 (1931).

81. S. Arrhenius and A. Westgren. Z. Phys. Chem. 14, 66 (1931).

82. V. P. Tarasov. Bull. Moscow University No. 4, 105 (1947).

83. Ya. S. Umanskii, B. N. Finkel'shtein, M. E. Blanter, S. T. Kishkin, N. S. Fastov, and S. S. Gorelik. Physical Metallography (Metallurgy Press, 1955) p. 761.

84. A. Westgren and G. Phragmen. Trans. Faraday Soc. 25, 382 (1929).

85. S. Fagerberg and A. Westgren. Metallwirtschoft 14, Apr., 265 (1935).

86. C. S. Barret. Structure of Metals (Metallurgy Press, 1948) p. 676.

87. F. R. Morral and A. Westgren. Arkiv. Kemi, Mineral Geol. 11B, 6 (1934).

88. G. Arrivaut. Z. anorg. Chem. 60, 436 (1908).

89. E. R. Jette and E. B. Gebert. J. Chem. Phys. 1, 753 (1933).

90. C. Di Capua. Att: accad. naz. Lincei 29, III (1920).

91. W. Loskiewicz. Przeglod Gorniczy – Hutniczy 21, 583 (1929); J. Inst. Met., London, 47, 516 (1931).

92. F. B. Litton and H. C. Andersen. J. Electrochem. Soc. 101, 287 (1954).

93. G. Masing and O. Dohl. Wiss. Veröffentl. Siemens-Koncern 8, 255 (1929).

94. H. A. Sloman. J. Inst. Met., London, 49, 369 (1932).

95. C. Winkler. Ber. 23, 2642 (1890).

96. P. Lebeau and B. Bossuet. Rev. met. 6, 273 (1909).

97. R. Vogel. Z. anorg. Chem. 61, 46 (1909).
98. L. Wöhler and O. Schliephake. Z. anorg. Chem. 151, 1 (1926).
99. W. Schmidt. Z. Metallk. 19, 452 (1927).
100. W. Mannchen. Z. Metallk. 23, 193 (1931).
101. E. A. Owen and G. D. Preston. Proc. Phys. Soc., London 36, 341 (1924).
102. W. Klemm and H. Westlinning. Z. anorg. Chem. 245, 365 (1941).
103. G. Busch and U. Winkler. Helv. Phys. Acta 26, 578 (1953).
104. R. F. Mehl, C. S. Barret, and F. N. Rhines. Trans. AIMME 203 (1932).
105. A. Guinier and H. Lambot. Compt. rend. 227, 74 (1948).
106. O. Kubaschewski and H. Villa. Z. Elektrochem. 53, 32 (1949).
107. H. Le Chatelier. Bull. soc. chim. 3, 17, 793 (1897).
108. L. Hackspill. Bull. soc. chim. 4, 3, 619 (1908).
109. A. Kolb. Z. anorg. Chem. 64, 342 (1909).
110. L. Wöhler and F. Muller. Z. anorg. Chem. 120, 49 (1921).
111. S. Tamaru. Z. anorg. Chem. 62, 81 (1909).
112. V. Louis and H. H. Frank. Z. anorg. Chem. 242, 117 (1939).
113. P. Eckerlin and F. Wölfel. Z. anorg. Chem. 280, 321 (1955).
114. E. Helner. Z. anorg. Chem. 261, 226 (1950).
115. H. Bohm and O. Hassel. Z. anorg. Chem. 160, 152 (1927).
116. H. Frank and V. Louis. Z. anorg. Chem. 242, 128 (1939).
117. W. Freundlich, A. Chretien, and M. Bichara. Compt. rend. 239, 1141 (1954).
118. L. Wöhler and W. Schuff. Z. anorg. Chem. 209, 33 (1932).
119. C. S. Bradley. Chem. News 82, 149 (1900).
120. E. Vigouroux. Compt. rend. 123, 115 (1896).
121. H. Moissan and F. Siemens. Ber. 37, 2086 (1904).
122. W. Guertler. Metallographie 1, 2, 702, Berlin (1917).
123. H. Moissan and A. Stock. Compt. rend. 131, 139 (1900).
124. F. Tone. Ind. Eng. Chem. 30, 236 (1938).
125. L. Brewer, D. L. Sawyer, D. H. Templeton, and C. H. Dauben. J. Am. Ceram. Soc. 34, 173 (1951).
126. C. S. Fuller and J. A. Ditzenberger. J. Appl. Phys. 25, 1439 (1950).
127. G. M. Samsonov and V. P. Latysheva. Proc. Acad. Sci. USSR 105, 499 (1955).
128. G. V. Samsonov and L. Ya. Markovskii. Prog. Chem. 25, 210 (1956).
129. W. Fraenkel. Z. anorg. Chem. 58, 154 (1908).

130. C. E. Roberts. J. Chem. Soc. 105, II, 1383 (1914).

131. A. G. Gwyer and H. W. Phillips. J. Inst. Met., London, 36, 294 (1926); 38, 31 (1927).

132. W. Broniewski and Smialowski. Rev. Met. 29, 542 (1932).

133. H. Spengler. Metall. 9, 181 (1955).

134. P. Ya. Sal'dau and M. V. Danilovich. Bull. Department of Physicochemical Analysis VI, 81 (1933).

135. V. G. Kuznetsov and E. S. Makarov. Bull. Department of Physicochemical Analysis 13, 177 (1940).

136. W. Klemm, L. Klemm, E. Hohmann, H. Volk, E. Orlamunder, and H. A. Klein. Z. anorg. Chem. 256, 239 (1948).

137. S. Tamaru. Z. anorg. Chem. 61, 44 (1909).

138. J. W. Nielsen and N. C. Baenziger. Acta Cryst. 7, 132 (1954).

139. C. D. Thurmond. J. Phys. Chem. 57, 827 (1953).

140. C. D. Thurmond and J. D. Struthers. J. Phys. Chem. 57, 831 (1953).

141. Schutzenberger. Compt. rend. 114, 1089 (1892).

142. Acheson. Chem. News 68, 179 (1893).

143. H. Moissan. Compt. rend. 117, 423, 425 (1893).

144. H. Moissan. Compt. rend. 140, 405 (1905).

145. A. M. Khantverger. Coll: "Abrasives" (United Sci. Tech. Press, Leningrad, 1935) p. 180.

146. J. Fitzgerald. "Carborundum", Monograff uber angew. (Electrochem., 1904).

147. Colson. Compt. rend. 94, 1316 (1882).

148. J. N. Pring. J. Chem. Soc. 93, 2101 (1908).

149. A. Taylor and D. Laider. Brit. J. Appl. Phys. 1, 174 (1950).

150. M. V. Kamentsev. Artificial Abrasive Materials (Mashgiz, Moscow, 1950).

151. A. Lampen. J. Am. Chem. Soc. 28, 851 (1906).

152. O. Ruff and M. Konschak. Z. Electrochem. 32, 515 (1926).

153. H. Nowotny, E. Parthe, R. Kieffer, and F. Benesowsky. Monatsh. 85, 259 (1954).

154. K. Arnt and E. Hausmann. Z. anorg. Chem. 215, 66 (1933).

155. G. Humphrey, S. Todd, J. Goughlin, and E. King. "Thermodynamics of carborundum", U.S. Bur. Mines Repts., Invest. No. 4888 (1952); C. A. 46, 8949 (1952).

156. H. Ott. Z. Krist., (A), 61, 515 (1925); 62, 201 (1925); 63, 1 (1926).

157. H. N. Baumann. J. Electrochem. Soc. 99, 109 (1952).

158. L. I. Ivanova. J. Gen. Chem. 21, 444 (1951).
159. R. Reuter and H. Knoll. Naturwiss. 34, 72 (1947).
160. G. Bormann and H. Seyfarth. Z. Krist. (A) 86, 472 (1933).
161. N. W. Thibault. Am. Mineralogist 29, 249, 327 (1944).
162. L. S. Ramsdell. Am. Mineralogist 30, 519 (1945).
163. G. S. Zhdanov and Z. V. Minervina. Proc. Acad. Sci. USSR 48, 192 (1945).
164. G. S. Zhdanov and Z. V. Minervina. J. Exp. Theo. Phys. 15, 655 (1945).
165. L. S. Ramsdell. Am. Mineralogist 32, 64 (1947).
166. G. S. Zhdanov and Z. V. Minervina. J. Exp. Theo. Phys. 17, 3 (1947).
167. G. Honjo, S. Miyake, and T. Tomita. Acta Cryst. 3, 396 (1950).
168. L. S. Ramsdell and J. A. Kohn. Acta Cryst. 4, 75 (1951).
169. E. B. Gasilova, M. S. Beletskii, and M. I. Sokhor. Proc. Acad. Sci. USSR 82, 57 (1952).
170. E. B. Gasilova and M. I. Sokhor. Proc. Acad. Sci. USSR 82, 249 (1952).
171. L. S. Ramsdell and J. A. Kohn. Acta Cryst. 5, 215 (1952).
172. L. S. Ramsdell and R. S. Mitchell. Am. Mineralogist 38, 56 (1953).
173. R. S. Mitchell. Am. Mineralogist 38, 60 (1953).
174. R. S. Mitchell. J. Chem. Phys. 22, 1977 (1954).
175. E. B. Gasilova. Proc. Acad. Sci. USSR 101, 671 (1955).
176. G. S. Zhdanov, Z. V. Minervina, and A. A. Nevzorova. Factory Labs. 14, 190 (1948).
177. G. B. Bokii. Introduction to Crystallochemistry (Moscow University Press, 1954) p. 343.
178. N. V. Belov. Structure of Ionic Crystals and Metallic Phases (Acad. Sci. USSR Press, Moscow, 1947) pp. 61-63.
179. G. Honjo. J. Phys. Soc. Japan, 4, 352 (1949); C. A. 44, 5698 (1950).
180. C. Kalb and W. Wittborg. Naturwiss. 38, 156 (1951).
181. A. R. Verma. Z. Electrochem. 56, 268 (1952).
182. H. E. Buckley. Z. Electrochem. 56, 275 (1952).
183. F. C. Frank. Phil. Mag. 7, 42, 1014 (1951).
184. H. Jagodzinski. Acta Cryst. 7, 300 (1954).
185. H. E. Merwin. J. Wash. Acad. Sci. 7, 445 (1917).
186. G. Busch. Helv. Phys. Acta 19, 167 (1946); 19, 189 (1946).
187. G. Busch and H. Labhart. Helv. Phys. Acta 19, 463 (1946).

188. G. Busch, P. Schmid, and H. Spondlin. Helv. Phys. Acta 20, 461 (1947).

189. G. Busch, H. Flury, and W. Mertz. Helv. Phys. Acta 21, 212 (1948).

190. R. L. Dudnik and V. M. Pruzhinina-Granovskaya. J. Theo. Phys. 19, 1434 (1949).

191. L. I. Ivanov and V. M. Pruzhinina-Granovskaya. J. Theo. Phys. 21, 1050 (1951).

192. W. Sasaki. J. Phys. Soc. Japan 7, 107 (1952).

193. J. T. Kendall. J. Chem. Phys. 21, 821 (1953).

194. M. B. Brodsky and D. Gubicciotti. J. Am. Chem. Soc. 73, 3497 (1951).

195. A. C. Lea. J. Soc. Glass Technol. 33, 27 (1949).

196. H. Salmang. Physik, und chem. Grundlagen der Keramik, (Spring.-Ver., Berlin, 1954) p. 275.

197. E. B. Gasilova, D. B. Gogoberidze, and S. V. Ukraintseva. Mineralogical Collection of the L'vov Geological Society, No. 7 29 (1953).

198. J. L. Everhard. Materials and Methods 34, 71 (1951).

199. G. I. Easter. J. Metals. 7, 805 (1955).

200. H. Stöhr and W. Klemm. Z. anorg. Chem. 241, 314 (1939).

201. F. X. Hassion, A. J. Goss, and F. A. Trumbore. J. Phys. Chem. 59, 1118 (1955).

202. A. Levitas. Phys. Rev. 99, 1810 (1955).

203. E. Vigouroux. Compt. rend. 123, 116 (1896).

204. S. Tamaru. Z. anorg. Chem. 61, 40 (1909).

205. O. Ruff. J. Electrochem. Soc. 68, 87 (1935).

206. M. B. Kamentsev. Proc. Acad. Sci. USSR 23, 665 (1939); J. Appl. Chem. 12, 1330 (1939).

207. A. N. Novikov. J. Appl. Chem. 20, 43 (1947).

208. R. Schwarz and T. Holfer. Z. anorg. Chem. 19, 321 (1925).

209. R. Schwarz. Ber. 55, 3242 (1922).

210. Magnus. Ann Phys. 4, 70, 303 (1923).

211. Mich, Immke, and Kratzers. Tonind. Ztg. 50, 1671, 1791 (1926).

212. Blix and Wirbelauer. Ber. 36, 4227 (1903).

213. E. Vigouroux. Ann. chim. phys. 7, 12, 43 (1897).

214. L. Weiss and T. Engelhardt. Z. anorg. Chem. 65, 38 (1910).

215. H. Funk. Z. anorg. Chem. 133, 67 (1924).

216. R. Schwarz and W. Sexauer. Ber. 59, 333 (1926).

217. E. Friedrich and L. Sittig. Z. anorg. Chem. 143, 293 (1925).

218. Matignon. Bull. soc. chim. 13, 791 (1913).

219. W. Roth and E. Borger. Ber. 70, 48 (1937).

220. Shun-Ichi-Satoh. Bull. Inst. Phys. Chem. Research (Tokyo) 1, 6, 51 (1937); Sci. Papers Inst. Phys. Chem. Research (Tokyo) 34, 144 (1938).

221. K. Kelly. Bull. U.S. Bur. Mines No. 407 (1937).

222. E. Weibk and O. Kubaschewsky. Thermochem. der Legierungen (Berlin, 1943).

223. W. B. Hincke and L. R. Brantley. J. Am. Chem. Soc. 52, 48 (1930).

224. W. C. Leslie, K. G. Carrol, and R. M. Fisher. J. Metals 4, 204 (1952).

225. J. E. Collins and R. W. Gerby. J. Metals 7, 612 (1952).

226. G. Shenk. Physicochemistry of Metallurgical Processes, Part I (United Sci. Tech. Press, Kharkhov, 1935) p. 384.

227. R. Inza and H. Hahn. Z. anorg. Chem. 244, 125 (1940).

228. W. Biltz. Sitzungsberichte der Preusischen (Akad. der Wiss. Phys. math. Klasse X, 1938) p. 99.

229. W. Klemm and P. Pirscher. Z. anorg. Chem. 247, 211 (1941).

230. R. S. Williams. Z. anorg. Chem. 55, 1 (1907).

231. R. Schwarz and H. Meckbach. Z. anorg. Chem. 232, 241 (1937).

232. R. Schwarz and R. Thiel. Z. anorg. Chem. 235, 247 (1938).

233. R. Schwarz and A. Köster. Z. anorg. Chem. 270, 2 (1952).

234. H. Schafer and J. Nickl. Z. anorg. Chem. 274, 250 (1953).

235. H. Schafer. Z. anorg. Chem. 274, 265 (1953).

236. I. W. Kurtschatow, T. Z. Kostina, and L. I. Rusinov. Phys. Z. Sowjetun. 7, 129 (1935).

237. B. M. Hochbery and M. S. Sominski. Phys. Z. Sowjetun. 13, 198 (1938).

238. P. F. Antipin and V. V. Sergeev. J. Appl. Chem. 27, 748 (1954).

239. G. Schott and W. Herrmann. Angew. Chem. 68, 213 (1956).

240. N. G. Klyuchnikov. J. Appl. Chem. 29, 130 (1956).

241. A. Chretien. Compt. rend. 241, 178 (1955).

242. V. Vand. Phil. Mag. 42, 1384 (1951).

243. A. Saulnier. Rev. Alumin. 32, 1011 (1955).

244. L. Pauling and P. Pauling. Acta Cryst. 9, 127 (1956).

245. F. Euer and W. Tromm. Z. anorg. Chem. 282, 29 (1955).

246. J. A. Stavrolakis, H. N. Barr, and H. H. Rice. Am. Ceram. Soc. Bull. 35, 47 (1956).

247. H. N. Patter. Trans. Am. Electrochem. Soc. 12, 191, 215, 223 (1907); German Pat. No. 189833, 1905; British Pat. No. 26788, 1906; French Pat. No. 360875, 1905; 366644, 1906.

248. R. Flusin. l'Ind. Chim. (Paris) 9, 391 (1922).

249. K. F. Bonhoeffer. Z. Phys. Chem. 131, 363 (1928).

250. C. A. Zapffe and C. E. Sims. Iron Age 149, 29, 34 (1942).

251. C. A. Zapffe. J. Am. Ceram. Soc. 27, 293 (1944).

252. H. Inuzuka. Mazda Research Bull. (Japan) 15, 305 (1940); C.A. 36, 4001 (1942).

263. W. Biltz and P. Ehrlich. Naturwiss. 26, 188 (1938).

254. E. Zintl, W. Brauning, H. L. Grube, W. Krings, and W. Morawietz. Z. anorg. Chem. 245, 1 (1940).

255. H. N. Baumann. Trans. Electrochem. Soc. 80, 95 (1941).

256. P. V. Gel'd and M. I. Kochnev. J. Appl. Chem. 21, 1249 (1948); Proc. Acad. Sci. USSR 61, 649 (1948).

257. P. V. Gel'd, A. G. Kologreeva, and N. N. Serebrennikov. J. Appl. Chem. 21, 1261 (1948).

258. P. V. Gel'd, O. A. Esin, N. N. Buinov, and R. M. Lerinman. Proc. Acad. Sci. USSR 67, 6 (1949).

259. P. V. Gel'd, A. I. Kholodov, and N. N. Buinov. Proc. Acad. Sci. USSR 70, 679 (1950).

260. P. V. Gel'd and O. A. Esin. Proc. Acad. Sci. USSR 70, 473 (1950).

261. P. V. Gel'd and N. N. Buinov. J. Appl. Chem. 23, 1087 (1950).

262. P. V. Gel'd and R. M. Lerinman. J. Appl. Chem. 23, 1191 (1950).

263. P. V. Gel'd and O. A. Esin. J. Appl. Chem. 23, 1200 (1950).

264. P. V. Gel'd and O. A. Esin. Trans. 4th Conference on Experimental Mineralogy 1, 154 (Acad. Sci. USSR Press, Moscow, 1951).

265. Gel'd and S. I. Popel'. J. Appl. Chem. 25, 465 (1952).

266. P. S. Mamykin, P. V. Gel'd, and N. N. Buinov. Proc. Acad. Sci. USSR 80, 701 (1951).

267. M. S. Beletskii and M. B. Rappoport. Proc. Acad. Sci. USSR 72, 699 (1950).

268. H. de Wet Erasmus and J. A. Persson. J. Electrochem. Soc. 95, 316 (1949).

269. G. Grube and H. Speidel. Z. Electrochem. 53, 339, 341 (1949).

270. H. Wartenberg. Z. Electrochem. 53, 343 (1949).

271. H. König. Optik 3, 419 (1948).

272. G. Hass. J. Am. Ceram. Soc. 33, 353 (1950).

273. M. Hoch and H. L. Jonston. J. Am. Chem. Soc. 75, 5224 (1953).

274. G. Jacobs. Compt. rend. 236, 1369 (1953).

275. N. C. Tombs and A. J. E. Welch. J. Iron Steel Inst. (London) 172, 69 (1952).

276. E. Gastinger. Naturwiss. 42, 95 (1955).

277. L. Brewer and R. K. Edwards. J. Phys. Chem. 58, 351 (1954).

278. S. Geller and C. D. Thurmond. J. Am. Chem. Soc. 77, 5285 (1955).

279. P. G. Saper. Phys. Rev. 42, 498 (1932).

280. W. Jevancs. Proc. Roy. Soc., London, 106, 174 (1932).

281. G. H. Wagner and A. N. Pines. Ind. Eng. Chem. 44, 321 (1952).

282. N. A. Shishakow. Nature 139, 927 (1937).

283. I. F. Ponomarev. Conferences of the All-Union Chemical Society 2, 21 (1946).

284. G. Seaborg, I. Perlman, and J. M. Hollander. Table of Isotopes [Russian Translation] (Foreign Lit. Press, Moscow, 1956) p. 27.

285. L. M. Frumina. Work Hygiene and Safety Practice 13, 81 (1935).

286. I. G. Ryss. Proc. Acad. Sci. USSR 24, 570 (1939); J. Phys. Chem. 14, 571 (1940).

287. Ya. Ya. Dodonov and M. I. Churmangeeva. J. Gen. Chem. 16, 1949 (1946).

288. A. P. Kreshkov and L. V. Anisimova. Trans. Mendeleev Chemico-Technological Institute, Moscow 12, 35 (Moscow, 1947).

289. N. V. Lazarev. Harmful Chemical Substances in Industry II (State Chem. Press, Moscow, 1951) p. 234, 253.

290. F. Flyuri and F. Tsernik. Harmful Gases (State United Sci. Tech. Press, Moscow, 1938) p. 300.

291. I. Moriya. Electrotech. J. 2, 219 (1938); C. A. 36, 3761 (1942).

292. C. N. Fenner. Am. J. Sci. 4, 36, 331 (1913).

293. W. Nieuwenkamp. Z. Krist. 90, 377 (1935).

294. W. H. Bragg and R. E. Gibbs. Proc. Roy. Soc., London, (A) 109, 405 (1925).

295. P'ei-Hsiu-Wei. Z. Krist. 92, 355 (1935).

296. E. Machatschki. Z. Krist. 94, 355 (1935).

297. R. E. Gibbs. Proc. Roy. Soc., London, (A) 110, 443 (1926).

298. R. W. Wyckoff. Z. Krist. 63, 507 (1926).

299. B. Gossner. Zentr. Mineral., (A) 329 (1927).

300. F. A. Hummel. J. Am. Ceram. Soc. 34, 235 (1951).

301. N. A. Shishakov. Proc. Acad. Sci. USSR 23, 791 (1939).

302. A. F. Frederickson. Am. Mineralogist 40, 1 (1955).

303. R. L. Myuller. J. Phys. Chem. 28, 1831 (1954).

304. H. Le Chatelier. Compt. rend. 108 (1046); 109, 264 (1889).

305. E. V. Tsinzerling. Proc. Acad. Sci. USSR 95, 529 (1954).

306. E. V. Tsinzerling. Trans. Institute Crystallography Acad. Sci. USSR 4, 209 (1948); 7, 81 (1952).

307. L. G. Chentsova and N. E. Vedeneeva. Trans. Institute Crystallography Acad. Sci. USSR 7, 159 (1952).

308. H. D. Keith. Am. Mineralogist 40, 530 (1955).

309. M. L. Keith and O. F. Tuttle. Experimental Investigations in the Field of Petrography [Russian Translation] (Foreign Lit. Press, Moscow, 1954) p. 156.

310. G. Sabatier. Compt. rend. 236, 720 (1953).

311. G. Sabatier and J. Wyart. Compt. rend. 239, 1053 (1954).

312. R. E. Gibson. J. Phys. Chem. 32, 1197 (1928).

313. H. S. Goder. Trans. Am. Geophys. Union 31, 827 (1950).

314. N. N. Sinel'nikov. Proc. Acad. Sci. USSR 92, 369 (1953).

315. H. Osterberg and J. W. Cookson. J. Franklin Inst. 220, 361 (1935).

316. H. Osterberg. Phys. Rev. 49, 552 (1936).

317. A. Pavlovic and R. Pepinsky. J. Appl. Phys. 25, 1344 (1954).

318. O. F. Tuttle and J. L. England. Bull. Geol. Soc. Am. 66, 149 (1955).

319. D. D. Eustachiv and S. Greenwald. Phys. Rev. 69, 532 (1946).

320. E. Buchler. Bell Labs. Record 31, 241 (1953).

321. A. C. Walker. J. Am. Ceram. Soc. 36, 250 (1953).

322. G. L. Humphry and E. G. King. J. Am. Chem. Soc. 74, 2041 (1952).

323. I. Simon and H. O. Memahon. J. Chem. Phys. 21, 23 (1953).

324. I. S. Kainarskii and L. I. Karyakin. Proc. Acad. Sci. USSR 81, 887 (1951).

325. A. Jayaraman. Proc. Indian Acad. Sci. 38A, 441 (1953); 48, 6773 (1954).

326. A. H. Jay. Proc. Roy. Soc. (A) 142, 237 (1933).

327. R. Evans. Introduction to Crystallochemistry (State Chem. Press, Moscow, 1948) p. 188.

328. R. W. Wyckoff. Am. J. Sci. 9, 448 (1925).

329. T. E. Barth. Am. J. Sci. 23, 35 (1932).

330. W. Johnson and K. W. Andrews. Trans. Brit. Ceram. Soc. 55, 227 (1956).

331. M. J. Buerger. Z. Krist. 90, 186 (1935).

332. W. Nieuwenkamp. Z. Krist. 92, 82 (1935); 96, 454 (1937).

333. R. E. Gibbs. Proc. Roy. Soc. (A) 110, 443 (1926).

334. T. E. W. Barth. Am. J. Sci. 24, 97 (1936).

335. D. S. Belyankin and B. V. Ivanov. Materials on the Study of Dinas and its Raw Material Base in the USSR (Acad. Sci. USSR Press, Moscow, 1938) p. 11.

336. A. H. Jay. Mineral. Mag. 27, 54 (1944).

337. E. Laves. Naturwiss. 27, 705 (1939).

338. D. S. Belyankin and N. G. Kaznakova. Trans. Petrographic Institute, Acad. Sci. USSR 6, 361 (1934).

339. I. S. Kainarskii and L. I. Karyakin. Proc. Acad. Sci. USSR 86, 137 (1952).

340. N. N. Sinel'nikov. Proc. Acad. Sci. USSR 102, 555 (1955).

341. H. E. Schwiete and H. Stollenwerk. Arch. Eisenhütten. 26, 583 (1955).

342. I. S. Kainarskii and E. V. Degtyareva. Proc. Acad. Sci. USSR 91, 355 (1953).

343. L. I. Karyakin and I. S. Kainarskii. Proc. Acad. Sci. USSR 86, 617 (1952).

344. A. I. Avgustinik and O. K. Krudevanidze. J. Appl. Chem. 19, 1189 (1946).

345. W. Büssem. Ber. deut. Keram. Ges. 16, 381 (1935).

346. G. Peyronel. Z. Krist. 104, 261 (1942).

347. I. S. Kainarskii and E. V. Degtyareva. Proc. Acad. Sci. USSR 99, 301 (1954).

348. R. B. Sosman. Am. J. Sci., Bowen Vol. 2, 517 (1952).

349. K. Khrushchev. Bull. Russian Academy of Sciences No. 1 (1895).

350. R. E. Gibbs. Proc. Roy. Soc. 113, 351 (1927).

351. E. Schiebold. Naturwiss. 18, 705 (1930).

352. M. A. Mosesman and K. S. Pitzer. J. Am. Chem. Soc. 63, 2348 (1948).

353. J. B. Austin. J. Am. Chem. Soc. 76, 6019 (1954).

354. R. B. Sosman. Trans. Brit. Ceram. Soc. 54, 655 (1955).

355. J. Lukesh. Am. Mineralogist, 27, 143 (1942).

356. T. E. Barth and A. Kvalheim. Norsk. Videnskaps-Akad. Oslo, 22, 1 (1944).

357. O. F. Florke. Naturwiss. 41, 371 (1954).

358. B. Mason. Am. Mineralogist 38, 866 (1953).

359. D. S. Belyankin and V. P. Petrov. Petrography of Georgia (Acad. Sci. USSR Press, Moscow, 1945).

360. V. I. Vlodavets. Collection devoted to Academician D. S. Belyankin (Acad. Sci. USSR Press, Moscow, 1946) p. 359.

361. S. N. Ruddlesden. Trans. Brit. Ceram. Soc. 54, 32 (1955).

362. C. W. Correns and G. Nagelschmidt. Z. Krist. A 85, 199 (1933).

363. A. Weiss and A. Weiss. Z. anorg. Chem. 276, 95 (1954).

364. F. Rinne. Z. Krist. A 60, 62 (1924).

365. F. Laves. Naturwiss. 27, 705 (1939).

366. P. P. Keat. Science 120, 328 (1954).

367. R. B. Sosman. Science 119, 738 (1954).

368. J. Endell. Kolloid Z. 111, 19 (1948).

369. L. Coes. Science 118, 131 (1953).

370. L. S. Ramsdell. Am. Mineralogist 40, 975 (1955).

371. R. W. Grimshaw, J. Hargreaves, and A. L. Roberts. Trans. Brit. Ceram. Soc. 55, 37 (1956).

372. G. W. Hull and T. H. Geballe. Phys. Rev. 96, 843 (1954).

373. T. Nemetschek. Z. Naturforschung. 11B, 148 (1956).

374. J. T. Law and E. E. Francois. J. Phys. Chem. 60, 353 (1956).

375. W. C. Dunlap, H. V. Bohm, and H. P. Mahon. Phys. Rev. 96, 822 (1954).

376. M. S. Maksimenko and A. S. Polubelova. Trans. Leningrad Technological Institute 33, 30 (1955).

377. L. I. Ivanov and V. I. Pruzhinina-Granovskaya. J. Theo. Phys. 26, 220 (1956).

378. L. S. Birks and J. H. Schulman. Am. Mineralogist 35, 8 (1950).

379. J. T. Randall, H. P. Rooksby, and B. S. Cooper. Z. Krist. 75, 196 (1930); J. Soc. Glass Technol. 14, 219 (1930).

380. L. Levin and E. Ott. Z. Krist. 85, 305 (1933).

381. N. N. Valenko and E. A. Porai-Kozhits. Nature 137, 273 (1936); Z. Krist. 95, 195 (1936).

382. H. R. Richter. Naturwiss. 40, 621 (1953).

383. N. A. Shishakov. J. Phys. Chem. 23, 889 (1949).

384. V. A. Florinskaya and R. S. Pechenkina. Proc. Acad. Sci. USSR 85, 1265 (1952).

385. N. A. Shishakov. Problems of the Structure of Silicate Glasses (Acad. Sci. USSR, Moscow, 1954).

386. V. A. Florinskaya and R. S. Pechenkina. Collection: The Structure of Glass (Acad. Sci. USSR Press, Moscow, 1955) p. 70.

387. S. P. Glagolev. Quartz Glass (State Chem. Press, Moscow, 1934) p. 216.

388. P. S. Narayanan. J. Ind. Inst. Sci. 35A, 9, 1953; C.A. 47, 5200, 1953.

389. V. P. Pryanishnikov. Collection: The Structure of Glass (Acad. Sci. USSR Press, Moscow, 1955) p. 270.

390. J. W. Marx and J. M. Sivertsen. J. Appl. Phys. 24, 81 (1953).

391. H. J. Mc Skimin. J. Appl. Phys. 24, 988 (1953).

392. M. E. Fine, H. Van Duyne, and N. T. Kennley. J. Appl. Phys. 25, 402 (1954).

393. O. L. Anderson and H. E. Bommel. J. Am. Ceram. Soc. 38, 125 (1955).

394. H. T. Smyth, H. S. Skogen, and W. B. Harsell. J. Am. Ceram. Soc. 36, 327 (1953).

395. H. D. Megaw. Z. Krist. 100, 58 (1938).

396. H. T. Smyth. J. Am. Ceram. Soc. 38, 140 (1955).

397. P. W. Bridgman. Am. J. Sci. 237, 7 (1939).

398. H. T. Smyth, J. W. Londeree, and G. E. Lorey. J. Am. Ceram. Soc. 36, 238 (1953).

399. P. W. Bridgman and I. Simon. J. Appl. Phys. 24, 405 (1953).

400. W. Primak, L. H. Fuchs, and P. Day. J. Am. Ceram. Soc. 38, 135 (1955).

401. F. Kracek. J. Am. Chem. Soc. 61, 2863 (1939).

402. A. W. Loubengayer and D. S. Morton. J. Am. Chem. Soc. 54, 2303 (1932).

403. D. M. Roy, R. Roy, and E. F. Osborn. J. Am. Ceram. Soc. 36, 185 (1953).

404. B. E. Warren and C. F. Hill. Z. Krist. (A) 89, 481 (1934).

405. W. R. Beck. J. Am. Ceram. Soc. 32, 147 (1949).

406. F. A. Hummel. J. Am. Ceram. Soc. 32, 320 (1949).

407. A. Dietzel and H. J. Poegel. Naturwiss. 40, 604 (1953).

408. A. Perloff. J. Am. Ceram. Soc. 39, 83 (1956).

409. M. J. Buerger. Am. Mineralogist. 39, 600 (1954).

410. P. P. Budnikov and E. I. Krech. J. Appl. Chem. 9, 1225 (1936).

411. F. Thummler and E. Klügel. Silikatechnik 4, 101 (1953).

412. J. Wyart and G. Sabatier. Compt. rend. 240, 1905 (1955).

413. G. W. Alexander, W. M. Heston, and R. K. Iler. J. Phys. Chem. 58, 453 (1954).

414. P. F. Holt and D. T. King. Nature 175, 515 (1955).

415. Heavens. Acta Cryst. 6, 571 (1953).

416. R. Mosebach. Neues Jahrb. Mineral., Abhandl. 87, 351 (1955).

417. G. W. Morey and J. M. Hesselgesser. Trans. ASME 73, 865 (1951).

418. G. W. Morey. J. Am. Ceram. Soc. 36, 279 (1953).

419. J. W. Gruner. Econ. Geol. 25, 697, 837 (1930).

420. L. Cambi. See Chem. Zentr. 11, 263 (1911).

421. W. Biltz. Z. anorg. Chem. 71, 82 (1911).

422. E. Tiede and M. Thimann. Ber. 59, II, B, 1703 (1926).

423. M. Picon. Compt. rend. 189, 97 (1929).

424. X. Sieber and E. J. Kohlmeyer. Arch. Erzbergbau, Erzaufbereit Metallhuttenw. I (2/3) (1930).

425. H. Gabriel and C. Alvarez-Tostado. J. Am. Chem. Soc. 74, 262 (1952).

426. E. J. Kohlmeyer and H. W. Retzlaff. Z. anorg. Chem. 261, 248 (1950).

427. W. C. Schumb and W. J. Bernard. J. Am. Chem. Soc. 77, 904 (1955).

428. C. R. Sabatier. Ann. chim. phys. (5) 22, 5 (1891).

429. O. Kubachewski and E. Evans. Thermochemistry in Metallurgy [Russian Translation] (Foreign Lit. Press, Moscow, 1954).

430. P. Rocquet and M. F. Ancey-Moret. Bull. chim. France, 1038 (1954).

431. A. Zintl and K. Loosen. Z. Phys. Chem. (A) 174, 301 (1935).

432. W. Büssem, H. Fischer, and E. Gruner. Naturwiss. 23, 740 (1935).

433. R. Schwarz. Z. anorg. Chem. 276, 33 (1954).

434. E. E. Vago and R. F. Barrow. Proc. Phys. Soc., London, 58, 538 (1946).

435. R. F. Barrow. Proc. Phys. Soc., London, 58, 606 (1946).

436. C. R. Sabotier. Compt. rend. 113, 132 (1891).

437. A. Weiss and A. Weiss. Z. Naturforschung 7b, 483 (1952).

438. A. Weiss and A. Weiss. Z. Naturforschung 8b, 104-105 (1953).

439. A. Weiss and A. Weiss. Z. Naturforschung 8b, 104 (1953).

440. A. Weiss and A. Weiss. Z. anorg. Chem. 273, 124 (1953).

441. D. K. Deardorff and E. T. Hayes. J. Metals 8; Trans. AIME 266, 509 (1956).

442. O. Hönigschmid. Compt. rend. 143, 224 (1906); Montasch. 27, 1069 (1906).

443. P. Askenazy and C. Ponnaz. Z. Elektrochem. 14, 810 (1908).

444. L. Baroduc and Müller. Rev. met. 7, 707, 748, 760 (1916).

445. P. P. Alexander. Met. a. All. 9, 179 (1938).
446. M. Hansen, H. Kessler, and D. McPherson. Trans. Am. Soc. Met. 44, 518.
447. C. M. Graigheat, O. W. Simmons, and L. W. Eastwood. Trans. AIME 188, 485 (1950).
448. D. I. McPherson and M. Hansen. Z. Metallk. 45, 76 (1954).
449. V. N. Eremenko. Titanium and Its Alloys (Acad. Sci. Ukrainian SSR Press, Kiev, 1955) p. 265.
450. P. Pietrakowsky and P. Duwex. J. Metals 3, 772 (1951).
451. F. Loves and H. L. Wallbaum. Z. Krist. 101, 78 (1939).
452. P. G. Cotter, J. A. Kohn, and R. A. Potter. J. Am. Ceram. Soc. 39, 11 (1956).
453. G. E. Hurdy and J. K. Hulm. Phys. Rev. 89, 884 (1953).
454. E. Wedekind. Ber. 35, 3932 (1902).
455. E. Wedekind. Z. Chem. and Ind. Kolloide 7, 249 (1900).
456. W. I. Kroll, A. W. Schlechten, W. R. Carmody, L. A. Jerkes, H. P. Helmes, and H. L. Gilbert. Trans. Electrochem. Soc. 92, 187 (1947).
457. C. E. Lundin, D. I. McPherson, and M. Hansen. Trans. Am. Soc. Metals 45, 901 (1953); for abridged translation, see: Collection: "Zirconium" No. II, 125 (Foreign Lit. Press, Moscow, 1955).
458. R. Kieffer, F. Benesowsky, and R. Machenschalk. Z. Metallk. 45, 493 (1954).
459. H. Schachner, H. Nowotny, and R. Machenschalk. Monatsh. 84, 677 (1933).
460. P. Pietrakowsky. Acta Cryst. 7, 435 (1954).
461. H. Schachner, H. Nowotny, and H. Kudielka. Monatsh. 85, 1140 (1954).
462. H. Seyfarth. Z. Krist. A 67, 295 (1928).
463. St. Nardy-Szabo. Z. Krist. A 97, 223 (1937).
464. G. Brauer and A. Mitius. Z. anorg. Chem. 249, 325 (1942).
465. G. L. Miller. Zirconium (Buttervorths Sci. Publ., London, 1954); [Russian Translation] (Foreign Lit. Press, Moscow, 1955).
466. B. Post, F. W. Glaser, and D. Moskowitz. J. Chem. Phys. 22, 1264 (1954).
467. R. Kiessling. Fortsch. Chem. Forsch. 3, 41 (1954).
468. H. Nowotny, H. Schachner, R. Kieffer, and F. Benesowsky. Monatsh. 84, 1 (1953).

469. H. Schachner, E. Cerwenka, and H. Nowotny. Monatsh. 85, 245 (1954).

470. E. Parthé, H. Schachner, and H. Nowotny. Monatsh. 86, 182 (1955).

471. E. Parthe, H. Nowotny, and H. Schmid. Monatsh. 86, 385 (1955).

472. E. Parthe, B. Lux, and H. Nowotny. Monatsh. 86, 859 (1955).

473. B. Aronsson. Acta Chem. Scand. 9, 137 (1955).

474. A. G. Knapton. Nature 175, 730 (1955).

475. H. Nowotny, E. Parthe, R. Kieffer, and F. Benesowsky. Monatsh. 85, 255 (1954).

476. F. Bertaut and P. Blum. Compt. rend. 236, 1055 (1953).

477. H. Moissan and Halt. Compt. rend. 135, 78, 493 (1902).

478. P. Lebeau. Ann. chim. phys. 8, 1, 553 (1904).

479. H. Giebelhausen. Z. anorg. Chem. 91, 251 (1915).

480. R. Vogel and C. Jenetzsch-Uschinski. Arch. Eisenhüttenw. 13, 403 (1940).

481. W. B. Pearson. J. Iron and Steel Inst. 164, 2, 155 (1950).

482. R. Kiffer, H. Schmid, and F. Benesowsky. Warmfeste u. korrosionsbest Sinterwerkstoffe (Spring.-Verl., Wien, 1956) p. 154.

483. H. J. Wallbaum. Z. Metallk. 31, 362 (1939).

484. H. J. Wallbaum. Z. Metallk. 33, 378 (1941).

485. G. W. Morey and E. Ingerson. Econ. Geol. 32, 607 (1937).

486. O. Hönigschmid. Monatsh. 28, 1017 (1907).

487. L. Brewer, A. W. Searcy, D. H. Templeton, and C. H. Dauben. J. Am. Ceram. Soc. 33, 291 (1950).

488. R. Kieffez, F. Benesowsky, H. Nowotny, and H. Schachner. Z. Metallk. 44, 242 (1953); see translation in the journal "Problems of Contemporary Metallurgy" 3, (2), 60 (1954).

489. G. C. Kennedy. Econ. Geol. 39, 25 (1944).

490. R. Kieffer and E. Cerwenka. Z. Metallk. 43, 101 (1952).

491. W. Meissner, H. Franz, and H. Westerhoff. Z. Physik 75, 521 (1932).

492. B. Aronsson. Acta chem. Scand. 9, 1107 (1955).

493. P. Lebeau and J. Figueras. Compt. rend. 136, 1329 (1903).

494. E. Vigouroux. Compt. rend. 144, 83 (1907).

495. G. DeChalmot. Am. Chem. J. 19, 69 (1897).

496. R. Frilley. Rev. met. 8, 468, 476, 484 (1911).

497. B. Boren. Ark. Kemi, Mineral. Geol. (A), 11, (10) 3, 1 (1933).

498. N. N. Kurnakov. Bull. Department of Physicochemical Analysis 16 (4) 77 (1948).

499. R. Kieffer, F. Benesowsky, and H. Schroth. Z. Metallk. 44, 437 (1953).

500. D. S. Bloom, J. W. Putnam, and N. J. Grant. J. Metals 4, Trans. 626 (1952).

501. K. Robinson. Acta Cryst. 6, 667, 854 (1953).

502. H. Moissan. Compt. rend. 120, 1320 (1895).

503. E. Vigouroux. Compt. rend. 129, 1238 (1899).

504. E. Defacqz. Compt. rend. 144, 1424 (1907).

505. J. L. Ham. Trans. AIME 73, 723 (1951).

506. R. Kieffer and E. Cerwenka. Z. Metallk. 43, 101 (1952).

507. D. H. Templeton and C. H. Dauben. Acta Cryst. 3, 261 (1950).

508. W. Zachariasen. Z. Phys. Chem. (B) 128, 39 (1927).

509. R. Kieffer. IVA 25, 264 (1954).

510. H. Schenk and N. Dehlinger. Acta Met. 4, 7 (1956).

511. F. W. Glaser. J. Appl. Phys. 22, 103 (1951).

512. M. J. Arvin. J. Appl. Phys. 24, 498 (1953).

513. T. B. Douglas and W. M. Logan. J. Research Nat. Bur. Standards 53, 91 (1954).

514. B. E. Walker, J. A. Grand, and R. R. Miller. J. Phys. Chem. 60, 231 (1956).

515. R. A. Long. Natl. Advisory Comm. Aeronaut. RME SOF 22, 34 (1950). Ceram. Abs. 35, 7 (1952).

516. I. E. Campbell, C. F. Powell, D. H. Nowicki, and B. W. Gonser. J. Electrochem. Soc. 96, 318 (1949).

517. R. Kieffer, F. Benesowsky, and C. Konopicky. Ber. deut. keram. Ges. 31, 223 (1954).

518. E. Fitzer and J. Schwab. Metall. 9, 1062 (1955).

519. E. A. Beidler, C. F. Powell, I. E. Campbell, and L. F. Intema. J. Electrochem. Soc. 98, 21 (1951).

520. E. Fitzer and Berg-und Hüttenmann. Monatsh. 97, 81 (1952).

521. E. Defacqz. Compt. rend. 144, 848 (1907).

522. R. Kieffer, F. Benesowsky, and E. Gallistl. Z. Metallk. 43, 284 (1952).

523. H. Nowotny, R. Kieffer, and H. Schachner. Monatsh. 83, 1243 (1952).

524. H. Nowotny, H. Schroth, R. Kieffer, and F. Benesowsky. Monatsh. 84, 579 (1953).

525. H. Nowotny, R. Machenschalk, R. Kieffer, and F. Benesowsky. Monatsh. 85, 241 (1954).

526. R. Kieffer and F. Benesowsky. Metall. 6, 171, 243 (1952).

527. D. J. Maykuth, H. R. Ogden, and R. I. Joffe. J. Metals 5, 231 (1953).

528. H. A. Wilhelm, O. N. Carlson, and J. M. Dickinson. J. Metals 6, 915 (1954).

529. O. Kubaschewski and H. Speidel. J. Inst. Metals 75, 417 (1949).

530. I. I. Kornilov and V. S. Mikheev. Prog. Chem. 22, 87 (1953).

531. E. Vigouroux. Compt. rend. 121, 771 (1895).

532. A. Carnot and G. Goutal. Ann. mines 18 (9), 271 (1900).

533. G. De Chalmot. J. Am. Chem. Soc. 18, 536 (1896).

534. P. Lebeau. Compt. rend. 136, 231 (1903).

535. F. Doerinckel. Z. anorg. Chem. 50, 117 (1906).

536. R. Vogel and H. Bedarf. Arch. Eisenhüttenw. 7, 423 (1933/34).

537. F. Laves. Z. Krist. 89, 189 (1934).

538. A. Westgren and G. Phragmen. Z. Physik. 33, 785 (1925).

539. C. S. Smith. Trans. AIMME Inst. Metals Div. 164 (1930).

540. A. W. Searcy and R. A. McNees. J. Am. Chem. Soc. 75, 1578 (1953).

541. R. A. McNees and A. W. Searcy. J. Am. Chem. Soc. 77, 5290 (1955).

542. N. V. Ageev and L. N. Guseva. Bull. Acad. Sci. USSR, Div. Chem. Sci. (1), 31 (1952).

543. N. V. Ageev. Bull. Department of Physicochemical Analysis 27, 75 (1956).

544. V. P. Elyutin, Yu. A. Pavlov, and B. E. Levin. Production of Iron Alloys (Metallurgy Press, Moscow, 1951) p. 496.

545. V. N. Verigin. Trans. All-Union Aluminum and Magnesium Institute No. 19 (1939).

546. V. P. Elyutin and R. N. Grigorash. Trans. Moscow Steel Institute 12 (1939).

547. A. A. Yaskevich and A. M. Samarin. Bull. Acad. Sci. USSR, Tech. Series No. 3 (1945).

548. G. De Chalmot. J. Am. Chem. Soc. 21, 58 (1899).

549. A. Weill. Rev. met. 42, 266 (1945).

550. P. Lebeau. Ann. Chem. Phys. (7) 26, 18 (1902).

551. W. Guertler. Metallographie 1, 658, 675 (1917).

552. F. Korber. Z. Elektrochem. 32, 371 (1926).

553. M. G. Corson. AIMME Tech. Publ. No. 96 (1928).

554. B. Stoughton and E. S. Greiner. AIMME Tech. Publ. No. 309 (1930).

555. N. S. Kurnakow and G. G. Urasow. Z. anorg. Chem. 123, 92 (1922).

556. J. L. Haughton and M. L. Becker. J. Iron Steel Inst. 121, 315 (1930).

557. G. Phragmen. Stahl O. Eisen 45, 299 (1925); 47, 193 (1927).

558. C. Krentzer. Z. Phys. 48, 558 (1928).

559. P. Oberhoffer and C. Kreutzer. Arch. Eisenhüttenw. 2, 449 (1929).

560. E. R. Jette and E. S. Greiner. Trans. AIMME Eng. Iron Steel Div. 105, 259 (1933).

561. N. Phragmen. J. Iron Steel Inst. 114, 397 (1926).

562. R. Wachtell. Trans. AIME 188, 354 (1950).

563. E. Fitzer. Z. Metallk. 44, 462 (1953).

564. F. Wever and H. Muller. Z. Krist. 75, 362 (1930).

565. N. V. Ageev, N. N. Kurnakov, L. N. Guseva, and O. K. Kononenko-Gracheva. Metallurgist 15 (1) 5 (1940).

566. N. S. Kurnakov and G. G. Urazov. Mining Journal (3), 167 (1914).

567. N. N. Serebrennikov and P. V. Gel'd. Proc. Acad. Sci. USSR 97, 695 (1954).

568. N. Kh. Abrikosov. Bull. Department of Physicochemical Analysis 27, 157 (1956).

569. T. Murakami. Sci. Repts. Tohoku Imp. Univ. 16, 475 (1927); see: Strukturbericht II, 759 (1937).

570. C. S. Smith and W. B. Daniels. Phys. Rev. 98, 1553 (1955).

571. J. A. Hedvall. Svensk. Kem. Tidskr. 37, 166 (1925); Refract. Bibliog. (Am. Ceram. Soc., Columbus, 1950) p. 738.

572. Ya. V. Dashevskii and S. I. Khitrin. Steel, No. 10, 892 (1948).

573. A. K. Prestrud. J. Iron Steel Inst. 169, 11, 107 (1951).

574. Ya. S. Dashevskii, A. E. Runov, I. S. Kozak, D. D. Zheltov, and B. A. Mel'nik. Steel 15, 714 (1955).

575. T. T. Khazanova and Yu. R. Vasin. Steel 15, 720 (1955).

576. Magnesital G. m. b. H.; French Pat. 902726, 1944; Cer. Abs. 31, 56 (1948).

577. W. L. Fink and K. R. Horn. AIMME, Tech. Publ. 351 (1930); Trans. AIMME, Inst. Metals Div. 383 (1931).

578. N. N. Kurnakov. Proc. Acad. Sci. USSR 34, 6 (1942).

579. R. Vogel and W. Mässenhausen. Arch. Eisenhüttenw. 27, 143 (1956).

580. R. Vogel and W. Schlüter. Arch. Eisenhüttenw. 12, 207 (1938).

581. K. Lewkonja. Z. anorg. Chem. 59, 327 (1908).

582. W. Guertler and G. Tammann. Z. anorg. Chem. 49, 93 (1906).

583. R. Vogel and K. Rosenthal. Arch. Eisenhüttenw. 7, 689 (1934).

584. B. Boren, S. Stahl, and A. Westgren. Z. Phys. Chem. (B) 29, 231 (1935).

585. S. Geller and V. M. Wolontis. Acta Cryst. 8, 83 (1955).

586. L. Pauling and A. M. Soldate. Acta Cryst. 1, 212 (1948).

587. H. Pfister and K. Schubert. Naturwiss. 37, 112 (1950).

588. F. Körber. Stahl O. Eisen 56, 1401 (1936).

589. W. Oelsen and Middel. Mitt. Kaiser-Wilhelm. Inst. Eisenforsch. 19, 1 (1937).

590. H. M. Greenhouse, R. F. Stoops, and T. S. Shevlin. J. Am. Ceram. Soc. 37, 203 (1954).

591. A. Kussmann and B. Scharnow. Z. Metallk. 23, 216 (1931).

592. O. Dahl and N. Schwartz. Metallwirtschaft 11, 277 (1932).

593. O. Dahl. Z. Metallk. 24, 277 (1932).

594. A. Osawa and M. Okamoto. Sci. Rep. Tohoku Univer. I. 27, 326 (1939).

595. K. Toman. Acta Cryst. 5, 329 (1952).

596. K. Toman. Acta Cryst. 4, 462 (1951).

597. W. Oelsen, Samson, and Himmelstjerna. Mitt. Kaiser-Wilhelm Inst. Eisenforsch. 18, 131 (1936).

598. H. Moissan and W. Manchot. Compt. rend. 137, 229 (1903); Ber. deut. chem. Ges. 36, 2993 (1903).

599. J. H. Buddery and A. J. E. Welch. Nature 167, 362 (1951).

600. S. Geller and E. A. Wood. Acta Cryst. 7, 441 (1954).

601. P. Lebeau and P. Jolibois. Compt. rend. 146, 1028 (1908).

602. A. T. Grigor'ev, T. A. Strunina, and A. S. Adamova. Bull. Platinum Department, Acad. Sci. USSR Press 27, 219 (1952).

603. H. Pfisterer and K. Schubert. Z. Metallk. 41, 358 (1950).

604. K. Schubert and K. Anderko. Naturwiss. 39, 351 (1952).

605. K. Anderko and K. Schubert. Z. Metallk. 44, 307 (1953).

606. P. Lebeau and A. Novitzky. Compt. rend. 145, 241 (1907).

607. E. Vigouroux. Compt. rend. 145, 376 (1907).

608. N. M. Voronov. Bull. Platinum Department, Acad. Sci. USSR Press 13, 145 (1936).

609. R. E. Carter and F. G. Richardson. Research, London, 7, 3 (1954).

610. W. H. Zachariasen. Acta Cryst. 2, 94 (1949).

611. F. Bertaut and P. Blum. Acta Cryst. 3, 319 (1950).

612. G. Brauer and H. Haag. Naturwiss. 37, 210 (1950).

613. G. Brauer and H. Haag. Z. anorg. Chem. 267, 198 (1952).

614. Sterba. Compt. rend. 135, 170 (1902).

615. R. Vogel. Z. anorg. Chem. 84, 323 (1913).

616. S. Smirnov-Verin. "Technique News", No. 23, 20 (1940).

617. H. Nowotny. Radex Rundschau (2) 41 (1953); Ceram. Abstr. 36, 169 (1953).

618. J. J. Harwood. Proc. Basic Mater. Conf. held in conjunction with the First Basic Mater. Exposit., N.Y., June, 1953; (Clapp and Poliak, N.Y., 1953) p. 27.

619. L. S. Ramsdell and J. A. Kohn. Acta Cryst. 4, 111 (1951).

620. A. F. Kapustinskii. J. Crystallography 1, 382 (1956).

621. O. Hönigschmid. Monatsh. 27, 205 (1906).

622. O. Hönigschmid, E. Wedekind, and K. Fetzer. Chem. Zentr. 29, 1031 (1905).

623. G. Brauer and A. Mitius. Z. anorg. Chem. 249, 325 (1942).

624. D. A. Robins and I. Jenkins. Acta Met. 3, 598 (1955).

625. E. Defacqz. Compt. rend. 147, 1050 (1908).

626. G. Brauer and H. Haag. Z. anorg. Chem. 269, 197 (1949).

627. J. J. Katz and E. Rabinowitch. The Chemistry of Uranium, Part I (McGraw-Hill Book Co., Inc., N.Y., 1951) p. 609.

628. Ya. S. Umanskii and T. V. Samsonov. J. Phys. Chem. 30, 1526 (1956).

629. G. T. Seaborg and J. J. Katz. The Actinide Elements (N.Y., 1954).

630. I. Sheft and S. Fried. J. Am. Chem. Soc. 75, 1236 (1953).

631. O. J. C. Runnalls and R. R. Boucher. Acta Cryst. 8, 592 (1955).

632. B. R. Frost and J. T. Maskrey. J. Inst. Metals 8, 177 (1953).

633. E. M. Savitskii and V. V. Baron. Bull. Department of Physico-chemical Analysis, Acad. Sci. USSR Press 27, 86 (1956).

634. H. Nowotny, B. Lux, and H. Kudielka. Monatsh. 87, 447 (1956).

635. H. Nowotny and H. Kudielka. Monatsh. 87, 471 (1956).

636. D. J. Maykuth, H. R. Ogden, and R. I. Jaffee. J. Metals 5, 231 (1953).

637. G. Phragmen. J. Inst. Metals 77 Part 6, 489 (1950).

638. W. Rostoker and A. Yamamota. Trans. Am. Soc. Metals 46, 1136 (1954).

639. C. H. Dauben, D. H. Templeton, and C. E. Myers. J. Phys. Chem. 60, 443 (1956).

640. H. Wartenberg. Z. anorg. Chem. 283, 372 (1956).

641. R. Meyer. Metaux 31, 219 (1956).

642. F. A. Wood. J. Phys. Chem. 60, 508 (1956).

643. N. N. Sinel'nikov. Proc. Acad. Sci. USSR 106, 870 (1956).

644. Metallurgia 53, 175 (1956).

645. G. Trömel and K. H. Obst. Arch. Eisenhüttenw. 26, 307 (1955).

646. O. W. Flörke. Ber. deut. Keram. Ges. 33, 319 (1956).

647. M. Schmeisser and M. Schwarzmann. Z. Naturforsch. 11b, 278 (1956).

648. W. B. Crandall and J. R. Tinklepaugh. Warmfeste u. Korrosionsbest. Sinterwerkstoffe (Spring.-Verl., Wien, 1956) p. 344.

649. E. L. Reed. J. Am. Ceram. Soc. 37, 146 (1954).

650. G. M. Carltan and L. S. Busch. U.S. Pat. No. 2747260 (1956); C.A. 50, 12432 (1956).

651. G. F. Hüttig and E. Härtl. Warmfeste u. Korrosionsbest. Sinterwerkstoffe (Spring.-Verl., Wien, 1956) p. 8.

652. E. Fitzer. Warmfeste u. Korrosionsbest. Sinterwerkstoffe (Spring.-Verl., Wien, 1956) p. 56.

653. E. A. Beidler and J. E. Campbell. U.S. Pat. No. 2665474. C.A. 48, 5780 (1954).

654. W. A. Maxwell. Materials and Methods 39, 200 (1954).

655. G. M. Ault and G. C. Deutsch. J. Metals 6, 1214 (1954).

656. British Pat. Notice 357/52; Metallurgia 45, 251 (1952).

657. Austrian Pat. No. 179100.

658. L. W. Coffer. U.S. Pat. No. 2622304.

659. J. E. Campbell, B. W. Gonser, and C. E. Powell. U.S. Pat. Nos. 2665475, 2665997, 2665998; C.A. 48, 5780 (1954).

660. G. A. Geach and F. O. Jones. Warmfeste u. Korrosionsbest. Sinterwerkstoffe (Spring.-Verl., Wien, 1956) p. 80.

661. H. Nowotny, H. Kudielka, and E. Parthe. Warmfeste u. Korrosionsbest. Sinterwerkstoffe (Spring.-Verl., Wien, 1956) p. 166.

662. D. A. Robins and J. Jenkins. Warmfeste u. Korrosionsbest. Sinterwerkstoffe (Spring.-Verl., Wien, 1956) p. 187.

663. I. G. Ryss. Chemistry of Fluorine and Its Inorganic Compounds (State Chem. Press, Moscow, 1956) p. 293.

664. R. Kieffer and F. Benesowsky. Iron Steel Inst., Sympos. on Powd. Met. 292 (1956).

665. J. P. Swentzel. U.S. Pat. No. 2752258 (1956).

666. H. M. Killmar and W. L. Wroten. Ceram. Ind. 66, 5, 93 (1956).

667. W. Rath. U.S. Pat. No. 2751188 (1956).

668. H. D. Erasmus and W. D. Forgeng. U.S. Pat. No. 2750268 (1956).

669. K. C. Nicholson. U.S. Pat. No. 2731359 (1956).

670. L. Brewer and H. Harabdsen. J. Electrochem. Soc. 102, 399 (1955).

671. H. A. Sloman. Metallurgia 53 (337), 135 (1956).

672. N. N. Murach. Chem. Sci. and Ind. 1, 492 (1956).

673. I. S. Kainarskii and I. G. Orlova. Collection: "Physicochemical Bases of Ceramics" (State Building Literature Press, Moscow, 1956) p. 507.

674. H. P. Rooksby and L. A. Thomas. Nature 149, 273 (1942).

675. I. S. Kainarskii and L. I. Karyakin. Proc. Acad. Sci. USSR 66, 1153 (1949).

676. E. T. Turkdogan and S. Ignatowicz. J. Iron Steel Inst. 185, Part 2, 200 (1957).

677. I. Simon. Phys. Rev. 103, 1587 (1956).

678. V. G. Zubov. Proc. Acad. Sci. USSR 107, 392 (1956).

679. A. I. Sarakhov. Bull. Acad. Sci. USSR, Div. Chem. Sci. No. 2, 150 (1956).

680. W. Störber. Kolloid-Z 145, 17 (1956).

681. R. Kieffer, F. Benesowsky, and H. Schmid. Z. Metallk. 47, 247 (1956).

682. L. N. Guseva and B. I. Ovechkin. Proc. Acad. Sci. USSR 112, 681 (1957).

683. F. W. Glaser and W. Iwanick. J. Metals 8; Trans. AIME 206, 1290 (1956).

684. W. Ptak. Hutnik 23, 234 (1956).

685. M. K. Disen and G. F. Huttig. Planseeber. Pulvermet. 4, 10 (1956); C. A. 51, 174 (1957).

686. J. A. Slyh, J. F. Lynch, and R. J. Runck. U.S. Pat. No. 2749254 (1956).

687. J. Slyh. Materials of the Atomic Energy Commission, USA, III, Nuclear Reactor Materials [Russian Translation] (Foreign Lit. Press, Moscow, 1956) p. 108.

688. R. A. Kempe and G. W. Croniger. U.S. Pat. No. 2763919 (1956).

689. L. Brewer and O. Krikorian. J. Electrochem. Soc. 103, 38, 701 (1956).

690. N. P. Zvereva. Collection: "Physicochemical Bases of Ceramics" (State Building Literature Press, Moscow, 1956) p. 325.

691. E. I. Salkovitz and F. M. Batchelder. J. Metals 4, 165 (1952).

692. G. White and S. B. Woods. Phys. Rev. 103, 569 (1956).

693. P. Lakodey, Ch. Euraud, and M. Prettre. Compt. rend. 242, 3071 (1956).

694. N. S. Gorbunov and A. S. Akopdzhanyan. J. Appl. Chem. 29, 655 (1956).

695. A. S. Akopdzhanyan and N. S. Gorbunov. J. Appl. Chem. 29, 659 (1956).

696. E. L. Jacobson, R. D. Freeman, A. G. Tharp, and A. W. Searcy. J. Am. Chem. Soc. 78, 4850 (1956).

697. S. M. Silverman. J. Chem. Phys. 25, 1081 (1956).

698. V. G. Zubov. Crystallography 1, 243 (1956).

699. I. A. Yakolev, L. F. Mikheeva, and T. S. Velichkina. Crystallography 1, 123 (1956).

700. H. Steinwehr. Z. Krist. B. 99, 293 (1938).

701. A. Perrier and R. Mandort. Compt. rend. 175, 622 (1922).

702. A. V. Shubnikov. Quartz and Its Use (Acad. Sci. USSR Press, Moscow, 1940).

703. J. F. MacDonald. Am. J. Sci. 254, 713 (1956).

704. A. Badaluco. La Ceramica 11, 3, 41; No. 6, 45; No. 11, 47 (1956).

705. M. Dank and S. Barber. J. Chem. Phys. 23, 597 (1955).

706. Tokyga. J. Jap. Ceram. Ass. 63, No. 715, 582 (1955); Referat. Zhur. No. 14, 292 (1956).

707. M. Wittels and F. A. Sherrill. Phys. Rev. 93, 1117 (1954).

708. A. Pflugmacher and J. Rohrmann. Z. anorg. Chem. 290, 101 (1957).

709. R. Schwarz and A. Pflugmacher. Ber. 75, 1062 (1942).

710. P. H. Keck and J. Broder. Phys. Rev. 90, 521 (1953).

711. F. A. Kurlyankin and N. A. Konovalova. Trans. Leningrad Technological Institute No. 34, 58 (1955).

712. N. F. Lashko. Proc. Acad. Sci. USSR 81, 605 (1951).

713. R. Ruttewit and G. Masing. Z. Metallk. 32, 52 (1940).

714. N. N. Zhuravlev. Crystallography 1, 666 (1956).

715. A. Kaufman, B. Cullity, and G. Bitsianes. J. Metals 9; Trans. AIME 209, 23 (1957).

716. N. A. Ageev and V. Samsonov. Proc. Acad. Sci. USSR 112, 853 (1957).